Mathematical Modelling

Modular Mathematics Series

Mathematical Modelling

J Berry

Centre for Teaching Mathematics
University of Plymouth

K Houston

Department of Mathematics
University of Ulster

ELSEVIER

Elsevier Ltd.
Linacre House, Jordan Hill, Oxford OX2 8DP
200 Wheeler Road, Burlington, MA 01803

Transferred to digital printing 2004

British Library Cataloguing in Publication Data
A catalogue record for this book is available from the British Library

ISBN 0 340 61404 8

1 2 3 4 5 95 96 97 98 99

Contents

Series Preface

This series is designed particularly, but not exclusively, for students reading degree programmes based on semester-long modules. Each text will cover the essential core of an area of mathematics and lay the foundation for further study in that area. Some texts may include more material than can be comfortably covered in a single module, the intention there being that the topics to be studied can be selected to meet the needs of the student. Historical contexts, real life situations, and linkages with other areas of mathematics and more advanced topics are included. Traditional worked examples and exercises are augmented by more open-ended exercises and tutorial problems suitable for group work or self-study. Where appropriate, the use of computer packages is encouraged. The first level texts assume only the A-level core curriculum.

Professor Chris D. Collinson
Dr Johnston Anderson
Mr Peter Holmes

Preface

'I have yet to see any problem, however complicated, which,
when looked at in the right way, did not become still more complicated'

Paul Anderson

This book is about the use of mathematics to solve problems in the real world.
Traditionally the discipline in which the use of mathematics is studied has been called
Applied Mathematics, but this term has often been associated with the application of
mathematics to science and engineering. Often, in school, Applied Mathematics is
associated with Mechanics, but mathematics occurs in many other subjects, for example
in economics, biology, linguistics, transport as well as in industry, commerce and
government.

Applying mathematics to such a wide range of subjects requires not only good
mathematical problem solving skills but the ability of the mathematician to start with a
problem in non-mathematical form and to give the results of any mathematical analysis
in non-mathematical form. In between these starting points are the important skills of
mathematical modelling. The process of mathematical modelling consists of three
main stages; we take a problem set in the real world and first *formulate* it as a
mathematical problem; this together with any assumptions made is the **mathematical
model**. The mathematical problem is then *solved* and finally the solution is translated
back into the original context so that the results produced by the model can be
interpreted and used to help solve the real problem.

To become skillful at mathematical modelling requires much hard work through
experience gained at problem solving. This book will help you to develop mathematical
modelling skills by solving problems; it is a "doing book" not a "watching book" in the
sense that it is important to have a go at some of the problems posed - you will not
become a good modeller by watching others.

The aim of this book is to develop your mathematical modelling skills in three different
ways:

- by solving simple modelling problems;
- by developing mathematical models in one application area, population growth;
- by reading and giving a critical appraisal of published articles.

The book has four chapters. Chapter 1 introduces the mathematical modelling process
through a wide range of real problem situations and answers the question 'what is
mathematical modelling?' Chapter 2 takes one particular application of mathematics to
population growth through which the mathematical modelling process is demonstrated.
Chapter 3 contains reprints of six articles from published books showing mathematical
modelling in action. The author of each article has developed one or more mathematical
models to describe a particular situation and has used the models to answer questions
about the situation. The aim of this chapter is to encourage you to see the modelling
process at work when you read mathematics. Chapter 4 contains activities to develop

your mathematics modelling skills further. Section 4.1 has more general problems to solve whereas Sections 4.2 and 4.3 introduce two alternative approaches to formulating models; simulation modelling and dimensional analysis modelling.

Study Guide

The book is written to support a first course in mathematical modelling. It is likely that not all the material will be used by every student on such a course. Our experience has shown that it is important in any modelling course to have a range of application areas for students from different backgrounds and interests. On our modelling courses we are often seeing students majoring in the sciences, social sciences, humanities as well as mathematics.

We would recommend that students study Chapters 1 and 2 fully as the material therein is designed to develop modelling skills as well as seeing the power of mathematics in modelling population growth. Subsequently, we would recommend that each student studies two of the articles from Chapter 3 and one section from Chapter 4; the choice depending on their main subject area. The modelling problems in section 4.1 can be used to consolidate the modelling skills developed in the earlier chapters; section 4.2 is probably suitable for students with a statistics background whereas Section 4.3 is for those with a mechanics or science background. We hope that Chapter 4 will provide a book for the diverse interest of students following modular type courses.

This book is the result of many years of teaching mathematical modelling to undergraduates and teachers on full-time and part-time courses at the Universities of Plymouth, Ulster and the Open University. Hence there are many colleagues in school, further and higher education with whom discussions over the past fifteen years have helped formulate our thoughts and courses in modelling. We would like to thank them and other people who has influenced the production of this book. Not wishing to upset those whose names we forget we have decided not to name any of these colleagues. But you know who you are! **So thanks**!!

We are grateful to the following publishers for permission to reproduce material from the following texts:

Open University Press, Course MST204 Unit 3, Figures 2(a)-2(n); Course M371 Block IV Unit 1;
Ellis Horwood, Modelling with Projectiles, Chapters 3 and 9; Applying Mathematics, Chapters 1, 18 and 19;
Oxford University Press, Mathematical Modelling, Chapter 7.

We thank Mrs Sharon Ward for turning our handwriting and our poor typing into the excellent CRC for this book. We appreciate her patience and advice in the many changes to the page design and layout. We also thank Michael Broughton for preparing the artwork.

To the reader, we wish *good modelling!*

John Berry
Ken Houston

1 •What is Mathematical Modelling?

There are three natural questions to ask at the beginning of this book

- why study mathematical modelling?
- what is a mathematical model?
- how do we find mathematical models?

Broadly speaking

- *mathematical modelling* provides a method for solving problems mathematically
- a *mathematical model* is a mathematical representation of the relationship between two or more variables relevant to a given situation or problem
- *finding mathematical models* is a skill that we hope you will develop in this course.

In this first chapter we begin to explore these questions and illustrate the answers through simple problem situations.

1.1 Modelling with data

One of the simplest *mathematical models* is a linear one which represents the relation between two variables by a straight line graph. In some cases the variables given in the problem satisfy a linear relation, but in other situations we might have to transform the variables to obtain a straight line graph.

For example, a common method of transformation is the use of logarithms. If the variables x and y satisfy a power law relation, $y = ab^x$, then a graph of y against x will produce a curve as shown in Fig 1.1a. However, taking logarithms of each side gives

$$\log y = \log a + x \log b$$

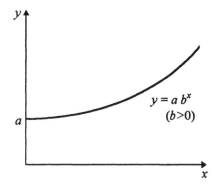

Fig 1.1a Graph of y against

and a graph of log y against x will give a straight line (as in Fig 1.1b). From the properties of the second graph we can estimate the values of a and b.

In the following two examples we illustrate the graphical approach to problem solving.

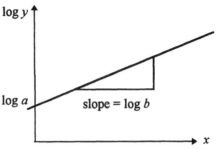

Fig 1.1b Graph of log y against x

Example 1 Modelling 'the greenhouse effect'

The burning of fossil fuels such as coal and oil adds carbon dioxide to the atmosphere around the earth. This may be partly removed by biological reactions, but the concentration of carbon dioxide is gradually increasing. This increase leads to a rise in the average temperature of the earth. Table 1.1 shows this temperature rise over the one hundred year period up to 1980.

Year	Temperature rise of the earth above the 1860 figure (°C)
1880	0.01
1896	0.02
1900	0.03
1910	0.04
1920	0.06
1930	0.08
1940	0.10
1950	0.13
1960	0.18
1970	0.24
1980	0.32

Table 1.1 The temperature rise of the earth over 100 years

If the average temperature of the earth rises by about another 6°C from the 1980 value this would have a dramatic effect on the polar ice caps, winter temperature etc. As the polar ice caps melt, there could be massive floods and a lot of land mass would be submerged. The UK would disappear except for the tops of the mountains!

Find a model of the above data and use it to predict when the earth's temperature will be 7°C above its 1860 value.

Solution

In this problem the variables are

• the temperature rise of the earth above the 1860 figure, T, and
• the year, n.

There is no simple way of discovering a relation between the temperature rise and the year by pure thought. There are many complicated processes going on in the atmosphere, and the effect on the atmosphere of burning fossil fuels will involve several physical laws and chemical reactions. However, we can make progress by representing the data graphically. It is probably quite clear to you that this set of data will not lead to a straight line graph and Fig 1.2 emphasises this point.

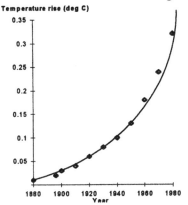

Fig 1.2 Graph of T against n

The graph of T against n is clearly not linear. However the points look close to a smooth curve passing through the origin. So we try to 'straighten the curve' by using logarithms. For example, if we plot log T against n we do indeed obtain a straight line graph through most of the data. The points at the lower end do not fit this straight line very well. (A simple explanation is that the data on temperature rise is correct to two decimal places. Clearly at the lower end the maximum percentage error in the data is much larger than that elsewhere. The temperature rise of 0.01 could be anything between 0.005 and 0.015 giving a maximum percentage error of 50%.

However the maximum percentage error for a temperature rise of 0.24 is 2.5%.

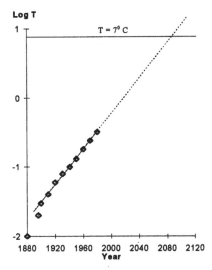

Fig 1.3 Graph of log T against n

From the graph in Fig 1.3 we can now predict when the temperature of the earth will be 7°C above the 1860 value. Drawing a horizontal line through the value log 7, we find a value for n as 2078. What we are saying is that if this linear relationship between $\log_{10} T$ and n is value for values of n outside the given range then in approximately 80 years time much of the UK will be flooded.

A sobering thought! Worried? Well you should be!

If you are 20 years old then what happens in 80 years time may be of little consequence; however, it will affect the children and grandchildren of the next generation. And of course the flooding will not suddenly happen but will occur over a period of many years before 2078.

But let's look at another problem that uses similar mathematical problem solving skills.

Example 2   World Record for the Mile

Table 1.2 shows the world record for the mile in minutes and seconds between 1913 and 1986

Time	Name	Country	Date	Place
4:14.4	John Paul Jones	USA	31.5.1913	Cambridge, Mass.
4:12.6	Norman Taber	USA	16.7.1915	Cambridge, Mass.
4:10.4	Paavo Nurmi	FIN	23.8.1923	Stockholm
4:09.2	Jules Ladoumegue	FRA	4.10.1931	Paris
4:07.6	Jack Lovelock	NZL	15.7.1933	Princeton, N.J.
4:0.6.8	Glen Cunningham	USA	16.6.1934	Princeton, N.J.
4:0.6.4	Sydney Wooderson	GBR	28.8.1937	Motspur Park
4:06.2	Gunder Hagg	SWE	1.7.1942	Gothenburg
4:06.2	Arne Andersson	SWE	10.7.1942	Stockholm
4:04.6	Gunder Hagg	SWE	4.9.1942	Stockholm
4:02.6	Arne Andersson	SWE	1.7.1943	Gothenburg
4:01.6	Arne Andersson	SWE	18.7.1944	Malmo
4:01.4	Gunder Hagg	SWE	17.7.1945	Malmo
3:59.4	Roger Bannister	GBR	6.5.1954	Oxford
3.58.0	John Landy	AUS	21.6.1954	Turku, Finland
3:57.2	Derek Ibbotson	GBR	19.7.1957	London
3:54.5	Herb Elliott	AUS	6.8.1958	Dublin
3:54.4	Peter Snell	NZL	27.1.1962	Wanganui
3:54.1	Peter Snell	NZL	17.11.1964	Auckland
3:53.6	Michel Jazy	FRA	9.6.1965	Rennes
3:51.3	Jim Ryun	USA	17.7.1966	Berkeley, Calif.
3:51.1	Jim Ryun	USA	23.6.1967	Bakersfield, Calif.
3:51.0	Filbert Bayi	TAN	17.5.1975	Kingston, Jamaica
3:49.4	John Walker	NZL	12.8.1975	Gothenburg
3:49.0	Seb Coe	GBR	17.7.1979	Oslo
3:46.31	Steve Cram	GBR	27.7.1985	Oslo
3:44.39	Noureddine Morceli	ALG	5.9.1993	Rieti

Table 1.2 The world record for the mile

Athletes continue to run the mile faster and faster as the years go by, but a mile in one minute, say, would seem to be impossible. Using the data estimate when it is likely that the mile could be run in 3 minutes 40 seconds.

Solution

If we draw a graph to represent the data in this problem then it is surprising how close to a straight line are the data points (see Fig 1.4)

Time in minutes for mile.

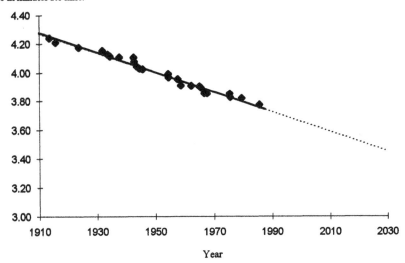

Fig 1.4 A graph of the world record for the mile

The graph is quite reasonably linear and from it we can predict that a mile run in 3 minutes 40 seconds could be achieved around the year 2000. We can continue the prediction process to suggest that a 3 minute 30 second mile will be run in about 30 years time in the year 2028. (To make these predictions we are assuming that the linear model holds true outside of the given data points).

We will have to wait to see how reasonable are these predictions. However Problems 1 and 2 suggest that a 3.5 minute mile will be achieved before the UK is flooded!

This method of solving real problems is fairly straightforward. We collect (or are given) data associated with a physical situation and use the data to draw a graph to represent the situation. Suppose that we label the variables x and y. We usually try one of three graphs in attempting to get a straight line between the variables:

- a graph of the variables, y against x, (for example Fig 1.4)
- a graph of the logarithm of one variable against the other (for example Fig 1.3)
- a graph of the logarithm of one variable ($\log y$) against the logarithm of the other ($\log x$)

Of course, if none of these leads to a straight line graph it does not mean that there is not a simple relationship between the variables. But finding the relationship may not be so easy.

6 Modelling with data

The graph that we have drawn is an example of **a mathematical model**. We can use the graph to obtain a relation in symbols. For instance in Example 2, the equation,

$$t = 255 - 0.38T,$$

where t is the time in seconds and T is the number of years beyond 1910 is a good fit to the straight line graph. Then the equation is another example of a mathematical model. In this way we are translating the problem from the real world situation to the rules and properties of the world of mathematics sometimes called **the mathematical world**. When the problem solving process is data driven as in Examples 1 and 2 we call the model an empirical model and the process of finding an **empirical model** is called **empirical modelling**.

TUTORIAL PROBLEM 1 Mathematical Models

List the mathematical structures that could form mathematical models.

TUTORIAL PROBLEM 2 Kepler's Third Law

Table 1.3 shows the distance of each of the planets from the sun (measured in millions of kilometres) and the time (measured in days) that it takes each planet to travel round the sun once, this time is called the period.

Planet	Distance from the sun $R(10^6)$ (km)	Period of revolution around the sun, T (days)
Mercury	57.9	88
Venus	108.2	225
Earth	149.6	365
Mars	227.9	687
Jupiter	778.3	4 329
Saturn	1427	10 753
Uranus	2870	30 660
Neptune	4497	60 150
Pluto	5907	90 670

Table 1.3 Planetary periods and distances from the sun

Use this data to find a mathematical model that relates the distance and the period. This is called Kepler's third law of motion for the planets, and was published in 1619.

In 1601, the German astronomer Johann Kepler became the Director of the Prague Observatory. After studying the motion of the planet Mars, in 1609 he formulated his first two laws for the motion of the planets:

1. each planet moves on a ellipse with the sun at one focus
2. for each planet, the line from the sun to the planets sweeps out equal areas in equal times.

TUTORIAL PROBLEM 3 Highway Code Distances

Table 1.4 shows the safe stopping distances for cars recommended by the British Highway Code.

Speed	Thinking distance		Braking distance		Overall stopping distance	
mph	m	ft	m	ft	m	ft
20	6	20	6	20	12	40
30	9	30	14	45	23	75
40	12	40	24	80	36	120
50	15	50	38	125	53	175
60	18	60	55	180	73	240
70	21	70	75	245	96	315

Table 1.4 Highway Code recommended stopping distances

Use these data to find mathematical models relating the various stopping distances and the speed of the vehicle.

The Highway Code gives certain conditions for the data to be appropriate. Find these conditions and discuss how your mathematical models are affected by them.

TUTORIAL PROBLEM 4 A leaking bottle

Take a plastic (see through) bottle and make a small hole near the base. Attach a vertical scale to the side of the bottle so that, when it contains water, you can read the height of the water surface in the bottle above the hole.

Cover the hole with your finger and fill the plastic bottle with water to the top of the scale. Remove your finger so that the bottle leaks water.

By collecting appropriate data, find an empirical mathematical model relating the height of water in the bottle above the hole and the time that has elapsed since the hole was uncovered.

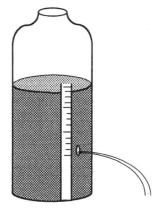

Fig 1.5 A leaking bottle experiment

TUTORIAL PROBLEM 5   Bode's Law

The solar system consists of the sun, the nine planets together with many smaller bodies such as the comets and the meteorites. The nine planets are, in order of distance from the sun, Mercury, Venus, Earth, Mars, Jupiter, Saturn, Uranus, Neptune and Pluto. Between Mars and Jupiter are many very small bodies called the minor planets or Asteroids and nearly all of them are too small to be seen by the naked eye from the earth. Tutorial Problem 2 was about Kepler's laws for the motion of the planets.

This problem is about an 18th century law called Bode's law and relates the distance of a planet from the sun to a number representing the planet. Table 1.5 gives the average distance of the planets from the sun and the ratio of these distances to the earth's distance from the sun.

Planet	Distance from the sun $R(10^6)$ (km)	Ratio R/R_e where R_e is the earth's distance from the sun
Mercury	57.9	0.39
Venus	108.2	0.72
Earth	149.6	1.00
Mars	227.9	1.52
Asteroids	433.8	2.9
Jupiter	778.3	5.20
Saturn	1427	9.54
Uranus	2870	19.2
Neptune	4497	30.1
Pluto	5907	39.5

Table 1.5 Data for Bode's law

Ignoring Mercury, we assign a number to each planet in the following way:

- for Venus choose n = 0
- for Earth choose n = 1
- .
- .
- .
- for Pluto choose n = 8

Use the data to find a model relating R/R_e and n. What value of n should you give to Mercury so that it too fits the model?

TUTORIAL PROBLEM 6   How much tape is left?

Most audio cassette recorders have a numerical tape counter which allows the user to create a numerical index for the items on the cassette tape for playback purposes. Furthermore, it is often convenient to be able to relate the number displayed on the tape counter with the playing time remaining, for example when needing to use the cassette to record a known length of music.

Does the counter on the machine operate so that the counter reading is directly proportional to the playing time, or does it count the number of revolutions of the take-up spool?

For example, for a C90 cassette (45 minutes playing time each time) the tape counter on my cassette deck goes from 0 to 600 for each side. Does it take 15 minutes to reach 200? When the counter reads 400 are there 15 minutes of playing time remaining?

For a cassette player equipped with a tape counter, formulate a mathematical model that describes the relationship between the counter reading and the amount of playing time that has elapsed.

TUTORIAL PROBLEM 7   The period of a pendulum

If you have seen a pendulum clock (for example a 'grandfather clock') you will realise that there must be a relationship between the period of swing and the effective length of the pendulum.

Set up a simple 'bob pendulum' using a heavy mass and a length of string. Measure the period of small swings for different lengths of the pendulum. Use your data to find a mathematical model that relates the period of the pendulum and the length of the string. Does the result depend on the mass of the bob?

Fig 1.6 A 'bob pendulum'

1.2 Using Mathematical Models

There are many types of problem that you will have tackled using mathematics. *Mathematical problems* are often set to develop and practice a mathematical skill. For example,

> find the two solutions of the quadratic equation
> $x^2 + 2x - 5 = 0$

is an example of a mathematical problem or 'exercise' which in this form has little relevance to any applications in the real world. *Mathematical investigations* are used to explore areas of mathematics which might be new to the learner. For example,

> Start with any 4 digit number where the digits are not all the same.
> Rearrange them in ascending and descending order and subtract the smaller from the larger.
> Repeat this with the number you obtain from the subtraction and continue. You will know when to stop.
> Try different starting numbers and investigate how long a chain of numbers you can obtain.

is an example of an investigation which for many provides an illustration of mathematics as a subject in its own right. *Real problems* are used to illustrate the powerful application of mathematics to solve problems set in the real world. For example in section 1.1 the problem concerning the 'greenhouse effect' is a problem about the world in which we live. We can use mathematics to attempt to solve many problems of this type. To distinguish between mathematical exercises, mathematical investigations and applications of mathematics to the real world we will call the latter **'real world problems'** and then mathematical modelling is the process for tackling such problems.

When using mathematics to solve real world problems one of our aims is to obtain a mathematical model that will **describe or represent** some aspect of the real situation. For example in Example 2, the world record for the mile was modelled by a graph and by an equation. Both of these models describe the relationship between the world record and the years between 1910 and 1990. Having found a mathematical model, we then use the model to **predict** something about the future. For example we were able to predict that a mile might be run in 3.5 minutes in the year 2028.

The formulation of a mathematical model can be a challenging task in many problems. Empirical models are fairly easy to find providing that we are given or can collect the data from appropriate experiments. But this problem solving process has severe limitations in the validity of our interpretations from the graph.

As an illustration of the caution needed when using an empirical modelling approach we consider Examples 1 and 2 again. Table 1.6 shows the predictions that we made in each case:

Problem	Predictions
Greenhouse effect	UK flooded in 2078
World Record for the mile	3 minute 40 sec mile in year 2002
	3 minute 30 sec mile in year 2028

Table 1.6 Predictions from Examples 1 and 2

At first sight these predictions may be reasonable, after all the curve through the data in Example 1 seems smooth and the straight line for Example 2 appears to fit the points well. If the predictions are reasonable then the 'greenhouse effect' should be a cause for concern. However, consider the world record for the mile. If we extend the line on the graph until it crosses the year axis then we can predict that 'by about 2582 the mile will be run in no time at all' i.e. the zero minute mile!! Clearly this method of approach is not very reliable. It is more likely that there is a curve through the data that gradually levels off to some limiting value and the straight line model only applies to a portion of the data.

So what concern should we have about the flooding of the UK? Unfortunately too many non mathematicians use this problem solving approach to cause alarm and to make false predictions. We should treat the solutions of Examples 1 and 2 with caution. Perhaps a 3.5 minute mile around the year 2028 is a reasonable prediction - it is not going too far beyond the range of the given data. However, the UK being flooded in the year 2078 is probably not a good prediction at all. In making this prediction we must assume that the rate of fossil fuel burning continues over the next hundred years as it has been over the past one hundred years. This is clearly unlikely. The production of electricity using different fuels to coal/oil is one example of changes that will occur. Industry and homes burn less coal and oil as the prices of these fuels have risen sharply during the past two decades. The motor car will eventually use a different polluting fuel that petroleum. It is interesting that in the mid-19th century there was concern that if the amount of horse manure dropped by horse drawn vehicles in London continued to increase at the then rate, the streets of London would be full of manure up the height of the highest building. This did not happen! Instead the motor car replaced the horse and a different pollution problem has occurred. So in one hundred years time the future generations may look back and smile at our concern over the 'greenhouse effect' - they will however have their own pollution problems!

To include changes in the important features of a problem in formulating our mathematical model requires a different problem solving process, one based more on theory than on data alone. This is called **theoretical modelling**. We introduce this in the next section.

The message of the first two sections is that, although formulating models (i.e. expressing relations between variables) using data is reasonably straightforward, and important as a problem solving tool, the method has severe limitations. We must list carefully the assumptions that are explicit to the model, and consider if and by how much, they may change before asserting the 'goodness' of any predictions.

Already in these two sections we have introduced three of the important skills in problem solving:

- understand the problem,
- be aware of the assumptions and simplifications made in solving the problem,
- question the results or predictions of the model.

The last skill is more than just saying "have I got the answer correct?" and "look in the back of the book"! The mathematical answer might be perfectly correct but the interpretation in the context of the real world is meaningless e.g. "a zero-minute mile will be run in the year 2582" is 'correct' mathematically but clearly nonsense physically.

TUTORIAL PROBLEM 8   Interpreting model predictions

Consider again the Tutorial Problems 2-7. List the assumptions and simplifications made in solving the problems and criticise the appropriateness of predictions.

TUTORIAL PROBLEM 9   Hooke's Law

You will need an elastic string or spring, masses and a support stand.
By hanging different masses from the ends of the elastic string or spring, formulate a mathematical model between the mass and the extension of the string or spring.

List the assumptions and simplifications made in formulating your model.
Discuss an improved model for the stretching of an elastic string or spring.

1.3 Modelling from Theory

In sections 1 and 2 we presented two examples showing the problem solving process based on formulating data-driven models and we discussed some of the drawbacks when making predictions based on this approach. The next three examples are intended to help you to gain an understanding of problem solving that uses a more theoretical approach. In each case notice how the data is used after the model is formed to help to check the validity of the model.

Example 3   The need for a pedestrian crossing

As a pedestrian there are many times in a day that you have to cross a road. For some roads, which do not carry much traffic you wait for a gap between cars and then cross; for more busy roads you are advised to cross at a zebra or pelican crossing. The local council has to decide whether and when to install controlled crossings on certain roads. This problem investigates road crossing strategies.

Formulate a mathematical model for crossing a one-way street so that a pedestrian can cross the road safely. Use your model to decide under what conditions a local council decides to install a pedestrian crossing.

SOLUTION

There are many factors which will affect the decision of the Local Council from cost through to the environment in which the road is part. Intuitively it is reasonable to think that the introduction of a pedestrian crossing will be dependent on the amount of road and pedestrian traffic. However other factors which are more difficult to quantify mathematically may have a more important bearing on the decision. For example, if the road is close to a hospital or a school then the decision to include a pedestrian crossing may not depend on the quantity of traffic but depends on the type of pedestrian traffic.

For a simple mathematical model consider the following assumptions and simplifications:

1. the road is one-way, a single carriageway and straight with no obstructions for the positioning of a pedestrian crossing,
2. the speed of the traffic is constant and equal to the road speed limit,
3. the density of the traffic is constant,
4. the pedestrians walk across the road at a constant speed.

The first statement ensures that when we have a simple mathematical model (based on assumptions 2, 3 and 4) the Local Council can install a pedestrian crossing without considering the environment. To formulate a mathematical model we need to introduce symbols to represent the physical quantities. The following table shows this information:

physical quantity	symbols	units
width of the road	w	metres
speed of the pedestrians	v	metres per second
time interval between the traffic	T	seconds

Table 1.7 The important variables in the problem

The time for a pedestrian to cross the road is w/v and the time between two road vehicles is T. A simple model in words is

> time to cross road < time between vehicles

and then in symbols we have

$$w/v < T$$

This is a condition for the pedestrians to cross the road safely. If we reverse the inequality then we have a simple condition for the need of a pedestrian crossing.

> Advice to the Local Council:
> Install a pedestrian crossing on a stretch of road if $w/v > T$.

To get a feel for the values in this inequality we need some data.

For the vehicle data suppose that the road is in a 30 mph (13.3 m/s) speed limit area and the vehicles travel at the safe distance recommended by the Highway Code (23 metres). With these values the time interval between the vehicles is 23/13.3 = 1.73 seconds.

For the pedestrians suppose that their speed is 4 mph (1.77 m/s) and the width of a single carriageway is 3 metres. The value of w/v is then 3/1.77 = 1.69 seconds.

With this data the advice to the council might be not to install a pedestrian crossing. However, the model is very simple and subject to many criticisms. For example

- the traffic is unlikely to be evenly spaced at the Highway code recommended distance;
- a safety margin of 0.04 seconds is not really realistic;
- the model suggests that you can either cross or not cross which is not realistic, the question of how long to wait does not enter the problem;
- 4 mph is quite a fast walking speed, especially for elderly people.

A more complicated and perhaps more realistic model is to assume that a pedestrian crossing will be installed when the probability of a gap of time interval greater than w/v is less than some predetermined value. The data for this model must show more detail than for the simple model above. Fig 1.8 shows the data collected by a group of students on a road in Plymouth during a thirty minute period.

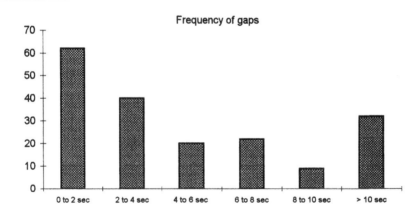

Fig 1.8 frequency distribution of time gaps between cars
on one carriageway of a Plymouth road

The probability of a gap of time interval greater than 2 seconds (this allows for a good safety margin) is calculated as the ratio

$$p = \frac{\text{number of gaps greater than 2 seconds}}{\text{total number of gaps}} = \frac{117}{180} = 0.65$$

The model says that the road needs a pedestrian crossing when the probability p is less than some predetermined value p_o.

The model could be developed further by considering

- the variability of the speed of the pedestrians;
- a statistical model involving the arrival times of cars and pedestrians.

There are several points to notice about this problem solving activity.

- The first model is very simple and straightforward, however it does allow us to obtain a better understanding of the problem and helps us to 'get into the problem'. An important saying in problem solving is *"a simple model is better than no model at all"*.
- In the first model there is a need for data to test or validate the model and in the improved model the data for a particular road is an important part of the formulation of the model.
- Each model that is formulated depends on certain assumptions and simplifications chosen by the problem solver. This allows different people to formulate different models and how good the models are can be tested at the validation stage with appropriate data.
- The mathematical model starts with a word description or **a word model** in each case. We then move from the words to the symbols which have been defined along with their units.
- The process of solving the problem is an iterative one; i.e. we start with a simple approach and them gradually refine it by looking back at the assumptions and simplifications.

Example 4   Icing Cakes

A wedding cake is to be baked in a square cake tin and will have a volume (before icing) of 4000 cm^3 . Determine the dimensions of the cake which will give the minimum surface area for icing (i.e. the top and the four sides).

Find the dimensions if the cake is baked in a circular tin.

There is a rule of thumb in cookery that one third of the marzipan should be used for the top of the cake and the remaining two thirds for the sides. Investigate the validity of this rule of thumb.

SOLUTION

The assumptions and simplifications for this problem are quite straightforward and will lead to the 'perfect cuboid or cylindrical cake'.

- The cake fills the tin exactly after baking and does not crumble or stick to the sides;
- each cake is perfectly flat on top and bottom so that it does not rise above the top of the cake tin;
- the mixture of volume 4000 cm^3 includes any air bubbles etc.;
- the marzipan goes on the cake before the icing.

1. Consider the square cake tin first. Let the sides of the cake have length x cm and the depth of the cake have length y cm.

   The volume of mixture in the tin is

   $$V = x^2 y = 4000$$

   and the surface area to be marzipanned and iced is

   $$S = x^2 + 4xy$$

   Eliminating y between these two equations gives

   $$S = x^2 + \frac{16000}{x}$$

   The mathematical problem is to find the value of x which gives the minimum value for S. There are several ways to do this, for example

   - by calculus, using differentiation;
   - drawing a graph using a calculator or spreadsheet.

   Whichever method you prefer, the minimum value of S ($= 1200$ cm^2) occurs when $x = 20$ cm and then $y = 10$ cm.

   The dimensions of a square cake with minimum surface area for icing is 20 cm × 20 cm × 10 cm.

2. Consider the circular cake tin. Let the radius of the cake be r cm and the depth of the cake be y cm.

   The volume of mixture in the tin is

   $$V = \pi r^2 y = 4000$$

   and the surface area to be marzipanned and iced is

$$S = \pi r^2 + 2\pi ry$$

Eliminating y between these two equations gives

$$S = \pi r^2 + \frac{8000}{r}$$

The minimum value of S is given by $r = y = 10.84$ cm (correct to 2 d.p.).

The test for the rule of thumb for the use of the marzipan is shown in the following table

shape of cake	area of top	area of sides	ratio
square	$20 \times 20 = 400$ cm^2	$4 \times (10 \times 20) = 800$ cm^2	2
circular	$\pi r^2 = 369.05$ cm^2	$2\pi ry = 738.11$ cm^2	2

Table 1.8 Testing the rule of thumb for icing a cake

So the 'rule of thumb' works for the minimum area of perfectly shaped cakes.

When testing this model two students at the University of Plymouth made the following observation:

> "The model really ought to be tested with real data to see if it was as accurate as we thought. Fortunately a cake tin of the size 20cmx20cmx10cm was found and a cake which had just been baked was used for the tests. On conclusion of the baking various things were found which tended to contradict the model. Firstly the cake had risen quite considerably in the middle and so was not flat on top - this of course changed the shape of the cake. Also the cake was rounded off on the corners and so was not perfectly square in shape. Also the cake contained a lot of air bubbles and was not smooth as we had assumed.
> When testing the rule of thumb it was found that the proportion of marzipan used on the top was in fact slightly greater than a third (more like 40%) although this was very difficult to measure."

The students have provided a good criticism of the model and the next stage would be to attempt to improve the model by including the feature that most cakes have a curved surface and are not perfectly flat.

Example 5

Most audio cassette recorders have a numerical tape counter which allows the user to create a numerical index for the items on the cassette tape for playback purposes. Furthermore, it is often convenient to be able to relate the number displayed on the tape counter with the playing time remaining, for example when needing to use the cassette to record a known length of music.

For a cassette player equipped with a tape counter, formulate a theoretical mathematical model that describes the relationship between the counter reading and the amount of playing time that has elapsed.

SOLUTION

In section 1.1 you solved this problem empirically by collecting data from an audio cassette recorder and drawing appropriate graphs. In this example we formulate a model from theory and then use the data to test the model.

To begin the formulation process we need to understand how the counter mechanism functions. Fig 1.9 shows a simple drawing of the mechanism of an audio cassette player. The tape leaves the supply spool, passes over the playback/recording heads at a constant speed and is collected up by the take-up spool. The constant speed is maintained by the capstan drive and pinch wheel. Since the tape speed across the heads needs to be constant the supply and take-up spools change speed during the playing (or recording) of a tape. We will assume that the tape counter is connected directly to the take-up spool.

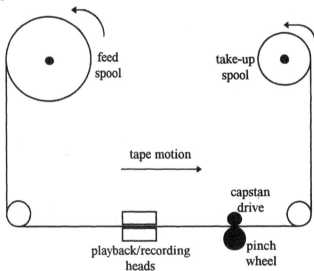

Fig 1.9 A simple view of the mechanism of an audio cassette player

From this introduction we deduce that the following features are important in formulating a mathematical model:

time elapsed
length of tape on the take-up spool
radius of the take-up spool when empty
radius of the take-up spool at general time
thickness of the tape
speed of the tape across the heads
counter reading
number of turns of the take-up spool
angle turned through by the take-up spool

As in the previous two examples the problem solving process begins with a list of assumptions and simplifications.

1. the speed of the tape across the heads is constant
2. the tape has constant thickness
3. the counter reading is a continuous variable
4. the counter reading number is proportional to the number of turns of the take-up spool

With these assumptions we can now proceed towards a mathematical model. In this example the model will consist of three sub-models for different parts of the system. This is a common feature of more complicated real problem solving activities. But first we define the variables, these are shown in the following table:

physical quantity	symbol	units
elapsed time	t (variable)	seconds
radius of the take-up spool at time t	r (variable)	cm
radius of the empty spool	r_0 (parameter)	cm
length of tape on the take-up spool	L (variable)	cm
thickness of the tape	h (parameter)	cm
counter reading	c (variable)	
angle turned through by the take-up spool at time t	a (variable)	radians
speed of the tape across the head	v (parameter)	cm/s

Table 1.9 The variables and parameters for the tape counter model

Note that in this table some of the quantities have been labelled as **variables** while others (r_0, h and v) are called **parameters** because they remain constant for a particular cassette player and tape; however if you change player and/or tape then the values of these quantities could change.

1. As the tape passes over the head at constant speed we have

    $$L = vt$$

2. Assumption 4 above states that the counter reading number is proportional to the number of turns of the take-up spool and since the number of turns is essentially the total angle turned through we have

$$c = ka$$

where k is a constant of proportionality.

Our aim is to formulate a model relating c
to t, so the next step is to find a relation
between L and a. To do this we consider the
length of tape added to the take-up spool δL
when it rotates through a small angle δa.

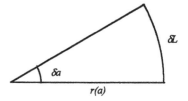

$$\delta L = r(a)\delta a \qquad\qquad (1)$$

Fig 1.10

Each time the take-up spool makes one complete revolution the angle a increases by 2π
and the radius of tape on the spool increases by the tape thickness h. So, using a
proportionate argument, if the angle increases by δa then the radius will increase by
$\delta r = h\delta a / 2\pi$. In the limit as δa tends to zero we have the differential equation

$$\frac{dr}{da} = \frac{h}{2\pi}$$

which has the solution

$$r = \frac{h}{2\pi}a + r_o \;.$$

Substituting in equation (1) and letting δa tend to zero we have

$$\frac{dL}{da} = \frac{h}{2\pi}a + r_0$$

Integrating (and using the initial condition $L = 0$ when $a = 0$)

$$L = \frac{h}{4\pi}a^2 + r_0 a$$

and finally substituting a by c/k and L by vt and rearranging we have

$$t = \frac{h}{4\pi vk^2}c^2 + \frac{r_0}{kv}c$$

This is a mathematical model relating the elapsed time t and the counter reading n.
Note that it is non linear which agrees with your empirical approach in Tutorial Problem
6.

To validate the model we need to use data collected from an audio cassette player. Table
1.10 shows the data from an experiment carried out by a group of students at Plymouth.
The radius of the spool was measured with a ruler and the constant of proportionality k
(for the equation $c = ka$) was calculated from the number of turns of the take-up spool
for the counter reading to increase by 100 (approximately 190 complete turns).

time t, (minutes)	counter reading, c	ratio t/c
0	000	
2	040	3.000
4	078	3.077
6	113	3.186
8	145	3.310
10	177	3.390
12	206	3.495
14	235	3.574
16	263	3.650
18	289	3.737
20	314	3.822
22	339	3.894

radius of the empty take-up spool $r_0 = 1.1$ cm
speed of tape across heads $v = 4.76$ cm/s
tape thickness $h = 0.0013$ cm
constant of proportionality $k = 0.084$ rad^{-1}

Table 1.10 Counter readings and elapsed time for an audio cassette player

We could draw a graph of t against c but it would not be as clear as a straight line graph; however if we divide each side of the model equation by c we obtain

$$\frac{t}{c} = \frac{h}{4\pi v k^2}c + \frac{r_0}{kv} \qquad (2)$$

This equation suggests that a graph of t/c against c should be a straight line. Fig 1.11 shows this graph for the data in Table 1.6 and a straight line looks a good fit.

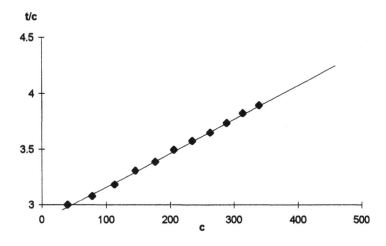

Fig 1.11 A graph of t/c against c for the tape data in Table 1.10

The values for the slope and intercept of the graph are 0.00315 and 2.832 respectively. From equation (2) and the values for h, v, k and r_0 we have

$$\frac{h}{4\pi vk^2} = 0.00309 \text{ (slope) and } \frac{r_0}{vk} = 2.76 \text{ (intercept)}.$$

The agreement between the model and the data is remarkably good. So we conclude that there is no need for an improvement to the mathematical model in this example.

List the important features that you think will be involved in formulating the model, list the assumptions and simplifications and criticise your model making appropriate suggestions for improvements.

TUTORIAL PROBLEM 10   Ropes and Knots

Tying a simple overhand knot in a rope will shorten it. Formulate a mathematical model between the shortening of the rope and its diameter. You should formulate your model without taking any measurements. However when you have a model then use ropes of different diameter to validate your model.

List the important features that you think will be involved in formulating the model, list the assumptions and simplifications and criticise your model making appropriate suggestions for improvements.

TUTORIAL PROBLEM 11   Length of a Toilet Roll

You are provided with a roll of toilet paper. Formulate a mathematical model to predict its length.

List the important features that you think will be involved in formulating the model, list the assumptions and simplifications and criticise your model making appropriate suggestions for improvements.

TUTORIAL PROBLEM 12   To Buy or Rent a TV

A person wishes to acquire a colour television set. The problem is "should the set be bought or rented?."

List the important features that you think will be involved in formulating the model, list the assumptions and simplifications and criticise your model making appropriate suggestions for improvements.

TUTORIAL PROBLEM 13 Second Hand Cars

The purchase of a new or second hand car is a major item of expenditure for most people. Some people always buy a new car, others always buy second hand cars. Again, some people change their cars frequently, say every two or three years, while others keep their cars for much longer, perhaps even when they are no longer roadworthy. By carefully considering the economics of car ownership one should be able to decide what is the best policy on changing one's car; whether to buy a new car or a second hand one, and in the latter case how old a car to buy and how long to keep it before trading it in.

Analyse the economics of car ownership, and decide on a strategy of how old a car should be when bought and for how long it should be kept.

List the important features that you think will be involved in formulating the model, list the assumptions and simplifications and criticise your model making appropriate suggestions for improvements.

TUTORIAL PROBLEM 14 Raffles

You are organising a raffle for a charity. The prizes will have to be paid for from the money raised by the sale of the tickets, and you have to decide the cost of the tickets and the value of the prizes. For a small prize, people would naturally only be willing to pay a small amount for a ticket, whereas for a large prize they would be willing to pay more.

Formulate a mathematical model to help you to decide the size of prize and the cost of the tickets.

List the important features that you think will be involved in formulating the model, list the assumptions and simplifications and criticise your model making appropriate suggestions for improvements.

Summary

This Chapter has introduced two approaches to problem solving. The first, leading to empirical models, relies on collecting (or using given) data and working from a graph. Although reasonably straightforward, and important as a tool, the method has severe limitations. We must list carefully the assumptions that are implicit to the model, and consider if, and by how much, they may change before asserting the "goodness" of any prediction.

In this section we have considered a more theoretical approach and you should notice how different the approach is to solving Examples 3 to 5 compared with Examples 1 and 2. The theoretical problem solving method is much more than just drawing graphs or solving equations. It contains the following steps:

- understand the problem
- identify the important features
- make assumptions and simplifications
- define variables
- use sub-models
- establish relationships between variables
- solve the equations
- interpret and validate the model (i.e. question the results of the model)
- make improvements to the model
- explain the outcome

The process of using these steps to solve a real world problem is called **mathematical modelling**. The process has essentially three phases; we **formulate** the mathematical model by describing or representing the real world in terms of mathematical structures (such as graphs, equations, inequalities, lising carefully the assumptions we make). We **solve** any equations that may occur (for example, in the tape recorder problem we needed to solve two differential equations). Then we use appropriate data to test the goodness of the model. In doing this we have **interpreted** the results of the mathematical analysis and criticised the model hopefully suggesting improvements to the model. This process is illustrated in Fig 1.12.

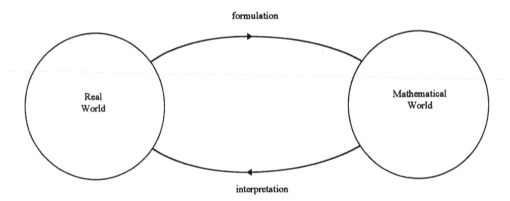

Fig 1.12 A simple view of Mathematical Modelling

In the next two chapters we develop this schematic view of modelling more fully.

2 •Modelling Population Growth

In Chapter 1 we introduced the main processes of empirical modelling and theoretical modelling. Developing good modelling skills can be done in two ways

- actually solving real problems
- working through the formulation of well known models.

You may have experience of using mathematics in science and in many ways we can think of science as being concerned with the establishment of mathematical models based on observation and experiment and the development and exploitation of new ideas based on the predictions of these models.

This chapter may be regarded as a case study in mathematical modelling. By constructing mathematical models for population growth we will go through the stages of modelling. Each section deals with a different stage and the summary puts the whole process together. The aim of this chapter is to see modelling at work through a study of population growth.

2.1 Understanding the Problem

The implications of an ever increasing world population cannot be overstated. The Royal Society has published a short report with stark facts:

we were 2 billion a few years ago,
we are 5 billion now, and
we will be 10 billion in fifty years time.

clearly such a growth rate is not sustainable with competition for space, natural resources, food, etc.

The need to model changes in the world population is important, but the process is not easy. There are many features to take into account, certainly too many for a first course in modelling and, having a model is not necessarily the answer to the problems. The modeller can make suggestions about birth rates, age distribution within a population, etc. but it does then take political decisions to attempt to solve the problems.

Prior to formulating mathematical models of population change we need to look at population data as well as the underlying concepts, features and variables that will make up the model.

Example 1   Population for England and Wales

The data in Table 2.1 shows the census data for England and Wales between 1801 and 1951. (There was no census in 1941).

Does the data show sufficient pattern to enable us to make predictions for future levels?

Solution

Perhaps the obvious observation is that the population is increasing and likely to continue to do so.

Fig 2.1 shows a graph of population against time

Year	Population
1801	8892536
1811	10164256
1821	12000236
1831	13896797
1841	15914148
1851	17927609
1861	20066224
1871	22712266
1881	25974439
1891	29002525
1901	32527843
1911	36070492
1921	37866699
1931	39952377
1951	43757888

Table 2.1

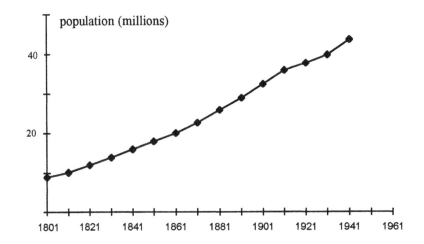

Fig 2.1 Population of England and Wales

The graph shows a steady growth in population throughout the period. The slope of the graph is called the **growth rate**. There is a clear decline in growth rate between 1911 and 1921 (the First World War obviously being a major influence).

The growth rate of a population is an important measure of how quickly the population is changing. Its value over a period of time can be estimated using the ratio

$$\text{average growth rate} = \frac{\text{change in population}}{\text{time interval}}.$$

For example, between 1801 and 1811 the average growth rate is 1.27×10^5 (people per year) and between 1901 and 1911 its value is 3.54×10^5 (people per year). So we could say that the growth rate has almost trebled in 100 years.

Now the average growth rate is a useful measure for comparing rates of change of a population at different times. However it is not such a good measure when comparing different populations. For example, suppose you are told that the average growth rate of England is one million per year and the average growth rate of Finland is one million per year. From this measure you have little idea of the effect of such a growth rate. However the population of England is ten times that of Finland (50 million compared with 5 million), so the same growth rate is likely to be more difficult for the Finnish people to cope with than for the English population.

A more useful measure of change in population is the **proportionate growth rate** defined by

$$\text{average proportionate growth rate} = \frac{\text{average growth rate}}{\text{population at start of the time interval}}.$$

So when comparing the (fictitious) growth rates of the populations of England and Finland their proportionate growth rates are $\frac{1}{50} = 0.02$ and $\frac{1}{5} = 0.2$ respectively and it is clear that Finland would have a bigger problem than England in absorbing such an increase.

Returning to the real data of Table 2.1, Table 2.2 shows the average proportionate growth rate for each ten year interval.

Interval start	1801	1811	1821	1831	1841	1851	1861
proportionate growth rate	0.143	0.181	0.158	0.145	0.127	0.119	0.121

Interval start	1871	1881	1891	1901	1911	1921
proportionate growth rate	0.144	0.117	0.122	0.109	0.050	0.055

Table 2.2 Proportionate growth rate for the population of England and Wales

The table shows a decreasing proportionate growth rate during the first half of the 19th century and with the exception of the 1870's a level proportionate growth rate during the latter half of the century; and clearly the drop in proportionate growth rate after 1911.

If we assumed the proportionate growth rate as constant at 0.055 (at 1921) continued from 1921 onwards the population of England and Wales would be as follows:

Year	1941	1951	1961	1971	1981	1991
Prediction	41949995	44047495	46249869	48562362	50990480	53540004

Table 2.3 Predicted population of England and Wales

TUTORIAL PROBLEM 1 Data Analysis

Figs 2.2(a)–(n) show data on a number of populations.
What trends can you propose from the data?

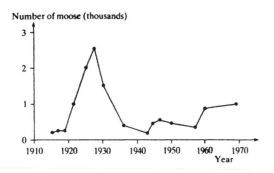

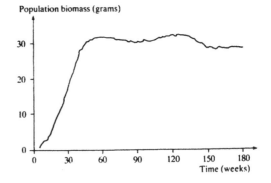

Fig 2.2(a) Estimates of the population of moose on Isle Royale in Lake Michigan. (Moose arrived on the island in about 1910).

Fig 2.2(b) Estimates of the number of wolves on Isle Royale. (Wolves arrived on the island in about 1948).

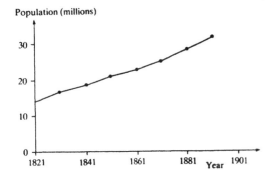

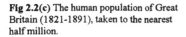

Fig 2.2(c) The human population of Great Britain (1821-1891), taken to the nearest half million.

Fig 2.2(d) The growth of a population of guppies in a laboratory aquarium (after Sulliman and Gutsell, 1958).

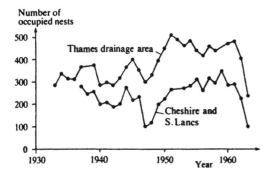

Fig 2.2(e) Number of breeding pairs of
heron (*Ardea cinerea*) in two parts of
England (1933-1963). Data from British
Trust for Ornithology, analysed by
J. Stafford (after Lack, 1966).

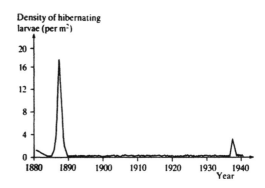

Fig 2.2(f) The population density of a moth
(*Dendrolimus*) in a forest at Letzlinger,
Germany.

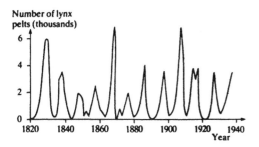

Fig 2.2(g) Variations in the populations of
lynx in the Hudson Bay area of Canada,
estimated from the number of pelts sold
to a company each year.

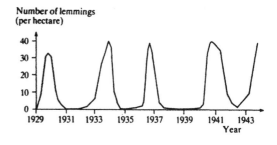

Fig 2.2(h) The population density of lemmings
in the area near Churchill, Canada.

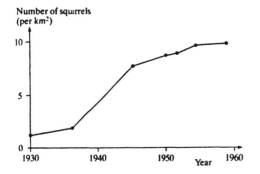

Fig 2.2(i) The population density of the grey
squirrel at selected sites in Great Britain.

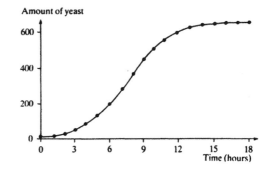

Fig 2.2(j) Growth of a laboratory culture of
yeast cells. Data from Carlson 1913 (after
Pearl, 1927).

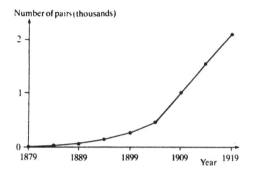

Fig 2.2(k) The population of gannets at a colony at Cape St. Mary in the St. Lawrence, Canada.

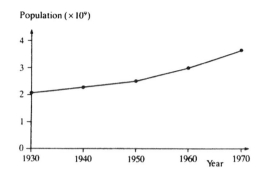

Fig 2.2(l) The human population of the world (1930-1970).

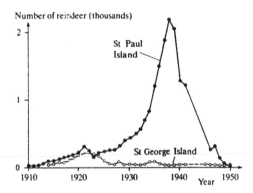

Fig 2.2(m) Reindeer population on two of the Pribilof Islands in the Bering Sea, from 1911, when they were introduced, until 1950. (After Scheffer, 1951).

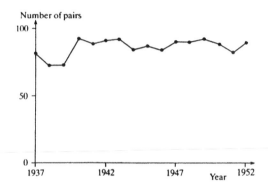

Fig 2.2(n) Population of the yellow-eyed penguin (*Megadyptes antipodes*) in Dunedin, New Zealand.

Notes: (i) In none of these examples is human exploitation believed to be a significant factor affecting population levels. (ii) Notice that various measures of the size of population are used. (iii) For wild populations, accurate counting can be difficult.

Certain of the data in these figures appear to suggest a constant population while others suggest a smooth change in population. You may feel fairly confident about your predictions in these cases.

However, for the data in Figs 2.2(e)-(h) quantitative predictions would be more difficult. We could say that the population would remain oscillating but it would be difficult to predict the size and period of the oscillations.

The aim of this section is to begin to formulate a theoretical model for population growth. This section is designed to give you practice in two important skills in mathematical modelling,

- identifying the important features of a problem;
- deciding which of the features should be included in the first simple mathematical model.

To formulate a theoretical model of some given situation or problem we need to choose variables and find relationships between them. A useful way of getting started on a complicated problem is

1. List as many features as you can which may have something to do with the problem or situation. Do not worry at this stage if you write down features that are irrelevant – sometimes an irrelevant feature may lead you to think of other more important features.

2. *Refine* your list to include relevant or 'sensible' features. At this stage it is useful to dismiss the irrelevant features and *sort* or *classify* those relevant features that are linked together and which influence each other.

Tutorial Problem 2 is about this process.

TUTORIAL PROBLEM 2   A feature list

Choose one of the species from the populations in Figs 2.2(a)-(n). Make a list of the features that you think need to be included in a mathematical model of the growth of population of a species. Group your features into appropriate classifications such as

(a)   features affecting births
(b)   features affecting deaths
(c)   features concerning the environment

etc.

2.2 Choosing Variables

Now that we have appropriate features for the problem or situation we can move towards a mathematical description.

Example 2

A rare species of bird is about to be introduced onto one of the Isles of Scilly. Formulate a mathematical model which describes the growth of the population of the species.

SOLUTION

Our feature list for the problem looks like this:

1.	initial number of birds	9.	population of predators
2.	size of the island	10.	diseases
3.	availability of food	11.	proneness of birds to fight
4.	numbers of males and females	12.	family size
5.	age distribution of birds	13.	role of unattached males
6.	life-span of birds	14.	will birds emigrate
7.	types of predators	15.	immigration of birds
8.	no. of eggs per female	16.	human exploitation

This list is refined and sorted into the following headings:

(a)   features affecting births
  1.   initial number of birds
  4.   numbers of males and females
  5.   age distribution of birds
  6.   life-span of birds
  8.   no. of eggs per female

(b)   features affecting deaths
  5.   age distribution of birds
  6.   life-span of birds
  7.   types of predators
  9.   population of predators
  10.   diseases
  11.   proneness of birds to fight
  12.   family size
  16.   human exploitation

(c)   features affecting food
  2.   size of the island
  3.   availability of food
  12.   family size

(d)   emigration and immigration
  13.   role of unattached males
  14.   will birds emigrate
  15.   immigration of birds

Now we take what might seem as a big step. Consider the four headings. They suggest the following equation in words:

$$\begin{pmatrix} \text{change in population} \\ \text{of birds during} \\ \text{a period of time} \end{pmatrix} = \begin{pmatrix} \text{number of} \\ \text{births} \end{pmatrix} - \begin{pmatrix} \text{number of} \\ \text{deaths} \end{pmatrix} + \left(\text{immigrants}\right) - \left(\text{emigrants}\right)$$

The features affecting food will affect the number of births and deaths and the ability of the island to sustain a large number of birds. Since the species of birds is rare it is likely that there will be no immigrants to the island in the early years. We will make a further simplification that there is no net emigration from the island i.e. all those birds that leave the island return the following year.

Suppose that P_n denotes the size of the population n years after the birds were introduced onto the island so that the increase in population in year $n + 1$ is $P_{n+1} - P_n$. Let B_n and D_n be the number of births and deaths respectively in year $n + 1$.

The word model can now be written in symbols as

$$P_{n+1} - P_n = B_n - D_n .$$

This is a simple mathematical model which assumes no immigration and no emigration:

TUTORIAL PROBLEM 3   Choosing Variables

Starting from your feature list in Tutorial Problem 2 form an appropriate word model for the problem of your choice. Choose appropriate variables to formulate a simple mathematical model.

2.3 Making Assumptions

The next step is to express the number of births and deaths in terms of population size and/or time. There could be many ways forward from here, but we'll keep it simple by making the following assumptions

1.   the annual number of births is proportional to the size of the population,

2.   the annual number of deaths is proportional to the size of the population.

In symbols we have

$$B_n = bP_n \quad \text{and} \quad D_n = dP_n$$

where b and d are constants, called the proportionate birth rate and the proportionate death rate respectively.

Hence the mathematical model for the bird population is

$$P_{n+1} - P_n = bP_n - dP_n = (b - d) P_n$$

The quantity $\dfrac{P_{n+1} - P_n}{P_n}$ is the proportionate growth rate introduced in section 2.1.

TUTORIAL PROBLEM 4

In the mathematical model formulated above, the quantity $(b - d)$ is constant.

Investigate the use of this model for the population of England and Wales introduced in Fig 2.1.

2.4   Solving the Equations

The model for population growth

$$P_{n+1} - P_n = (b - d)\, P_n$$

is an example of a recurrence relation. Numerical solutions can be found using a calculator or spreadsheet.

TUTORIAL PROBLEM 5

In May 1980 a small herd of 50 deer was introduced into a country park. Observations during the past few years suggest that each year the average breeding success of a female deer is 2 young reared to maturity, while, during each year 5% of the population at the start of the period die.

(a)   Use assumptions of the type in section 2.3 to formulate a mathematical model describing the population growth

(b)   Use a calculator or spreadsheet to find the size of the herd each year between 1980 and 1990.

(c)   How many deer should be killed each year so that the landowner can maintain a herd size of 200 deer.

(d)   Comment on the assumptions you have made and discuss if they need to be amended with time.

TUTORIAL PROBLEM 6   The General Solution

Show (by substitution) that the general solution of the recurrence relation model for population growth is

$$P_n = (1 + b - d)^n\, P_0$$

where P_0 is the population size at $n = 0$.

2.5 Interpreting the Solution and Validating the Model

The general power solution of the simple model with constant birth and death rates is called **the exponential model**. The solution gives a formula for the population size P_n in terms of time n. But how good is the model?

TUTORIAL PROBLEM 7 Interpreting the Solution

Sketch graphs of the general solution in the three cases

(a) $b - d > 0$,
(b) $b - d = 0$,
(c) $b - d < 0$.

Describe in words the long term behaviour of the population in each case. Comment on the applicability of the exponential model to real populations.

TUTORIAL PROBLEM 8 Validating the Model

The graph of $\log_e(P_n)$ against n should be a straight line for the exponential model. Explain why?

1.   Consider the census population for England and Wales between 1801 and 1901. Does this fit an exponential model?

2.   Look back at the population data given in Fig 2.2(a)-(n). Comment on the appropriateness of the exponential model as a good model in each case.

2.6 Criticising and Improving the Model

There are many ways that we can criticise the exponential model.

- if $b > d$ the population will increase in size forever;
- if $b < d$ the population tends to zero;
- the model predicts fractional species i.e. 'half a person';
- proportionate birth rate and proportionate death rate are independent of the size of the population and time and are taken to be constants;
- the balance of males and females and age distribution are not included;
- environmental conditions and changes are not included.

Now for some populations, and for limited periods, the exponential model appears to be a fairly good description of population change. In the early stages of development of a species we could argue that neglecting the composition of the population and the effects of the environment is a reasonable strategy. Provided we have a good mix of males and females, appropriate food, and space, then the population should increase unchecked as the exponential model predicts.

The 'fractional species' is less of a problem and will not have a significant effect on the model. It is a mathematical technicality in which the population size will be an integer whereas the mathematical variable representing the population may take any real number (including negative numbers!). Clearly we need to be careful to interpret the solutions sensibly.

Eventually the age distribution within the population and, as the population size grows, competition for food and space will become important features leading to the need to change the assumptions about constant birth and death rates.

TUTORIAL PROBLEM 9   Improving the Model

1.   Consider the data in Figs 2.2(a)-(n). Which curves would appear to be improvements of the exponential model?

2.   Suggest assumptions that incorporate effects of overcrowding in the population.

The growth curve which has features similar to the exponential model in the early years is called **the logistic curve**.

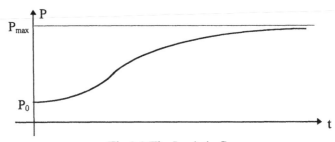

Fig 2.3 The Logistic Curve

The logistic curve increases like the exponential model for low population sizes but then tends towards a maximum sustainable population P_{max} say. This implies that when $P = P_{max}$ the species can exist at this level, there being sufficient food and space to sustain this population level.

The exponential model is

$$P_{n+1} - P_n = (b - d) P_n$$

where b and d are the proportionate birth and death rates which are assumed to be constant. We now relax this condition and assume that as the population grows in size then b and/or d are themselves functions of the population size, so that

$$b - d = f(P_n)$$

for some function f. A simple form for $f(P)$ is a decreasing linear function with $f(P) = 0$ when $P = P_{max}$ so that the change in population is zero at the maximum sustainable population.

With these assumptions $f(P)$ takes the form

$$f(P) = k\ (P_{max} - P)$$

for constant k. The recurrence relation model for population growth is

$$P_{n+1} - P_n = k\ P_n\ (P_{max} - P_n)\ .$$

This non-linear equation is called **the logistic model** and finding an analytical solution is more difficult. However we can interpret and validate the model using numerical solutions.

TUTORIAL PROBLEM 10 Investigating the Solution of the Logistic Model

(a)   Choose different values for k, P_0 and P_{max} and use a calculator, spreadsheet or computer package (such as DERIVE) to investigate solutions of the logistic model. Sketch graphs of your solutions for $P_0 > P_{max}$ and for $P_0 < P_{max}$.

(b)   Look back at Figs 2.2(a)-(n). Comment on the appropriateness of the logistic model as a good model in each case.

TUTORIAL PROBLEM 11 Validating the Logistic Model

Show that for the logistic model to be a good model for a particular set of population data the graph of $\dfrac{P_{n+1} - P_n}{P_n}$ against P_n should be a straight line.

For which of the following data is the logistic model a good model?

USA population data 1850-1910

(a)	Year	1850	1860	1870	1880	1890	1900	1910
	Population (millions)	23.2	31.4	38.6	50.3	62.9	76.0	92.0

Growth of a population of bacteria in laboratory conditions

(b)

Time (hours)	1	2	3	4	5	6
Population	1	6.63	16.8	41.8	97.6	201.9

Time (hours)	7	8	9	10	11	12
Population	344.3	473.4	553.7	592.5	609.1	615.7

Population of British Collared Doves (Hudson, 1965)

(c)

Year	1959	1960	1961	1962	1963	1964
Population	205	675	1900	4650	10200	18855

TUTORIAL PROBLEM 12   Criticising and Improving the Logistic Model

List the assumptions needed in the formulation of the logistic model. Consider the strengths and weaknesses of these assumptions. How well do the assumptions include the feature lists for population modelling of section 2.1?

Suggest possible improvements to the logistic model.

TUTORIAL PROBLEM 13   Differential Equation Models

(This problem is for those readers who have met calculus.)

The exponential and logistic models of population growth can be formulated in terms of differential equations by considering a small time interval h, say, and assuming that the number of births in the time interval equals birthrate $\times$ time interval and similarly for the number of deaths.

Formulate and investigate the solutions of the following differential equation models

$$\text{exponential model} \qquad \frac{dP}{dt} = (b-d)P$$

$$\text{logistic model} \qquad \frac{dP}{dt} = a\,P\left(1 - \frac{P}{P_{max}}\right)$$

Summary

In this chapter you have seen the process of mathematical modelling at work through population modelling. The key activities in the modelling process are summarised in the following list.

1. Understand the problem

    - decide what aspect of the problem to investigate
    - collect and analyse some data appropriate to the problem

2. Choosing variables

    - 'brainstorm' the situation or problem to form a feature list
    - refine and sort your list into the key features
    - for the key features define variables to be used in your model

3. Set up a mathematical model

    - try to describe the situation or problem as a word model
    - write your word model in symbols using the defined variables
    - state word model and mathematical model, remember that a simple model is easier to work with initially and a simple model can often bring an insight into the situation or problem that may help in your later improvements.

4. Formulate and solve the mathematical problem

    - often a mathematical modelling activity leads to the setting up and solution of a mathematical problem, at this stage you should keep to familiar mathematical territory.

5. Interpret the solution

    - describe the solution in words, is there qualitative agreement between the outcome of your model and the situation being considered
    - decide what data you need to validate your model and collect them.

6. Compare with reality

    - test the outcomes of your model with appropriate data
    - criticise your model, in particular, look back at your simplifying assumptions.

7.   Improve the model

- revise your simplifying assumptions
- formulate a revised model
- repeat the processes of solving, interpreting and validating.

8.   Report on your modelling activity

- prepare a report describing the problem and its outcomes, this might be in the form of a poster, a written report or an oral presentation.

The list of activities above is not a definitive list that you might always go through in order. You may want to hop about between different headings, this is often the way good modellers work. However if you get stuck in problem solving then the list often provides some guidance or tasks to try.

Note how the list suggests that mathematical modelling is an iterative process in that you are going round a loop, hopefully improving your model during each cycle. The following diagram emphasises the cyclic approach to modelling.

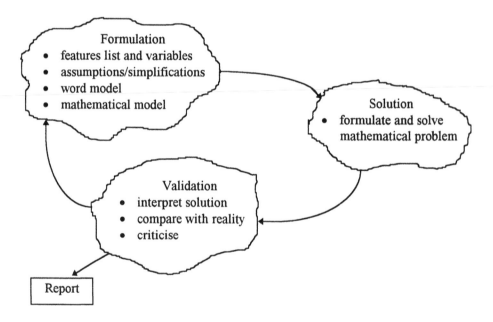

Fig 2.4 The modelling process

Exercises

1.   In a population of birds the proportionate birth rate and the proportionate death rate are both constant, being 0.45 per year and 0.65 per year respectively.

Immigration occurs at a constant rate of 2000 birds per year and emigration at a constant rate of 1000 per year.

(a)   Use these assumptions to formulate a recurrence relation to model the population.

(b)   Without solving the equation describe the long-term behaviour of the population in the two cases when the initial population is 3000 or 8000.

(c)   Use a calculator or computer and software package (such as a spreadsheet) to solve the recurrence relation to verify your results in (b).

2.   The population of fish in a reservoir is affected by both fishing and restocking. The proportionate birth rate is constant at 0.6 per year and the proportionate death rate is constant 0.65 per year. The reservoir is restocked at a constant rate of 4000 fish per year and fishermen are allowed to catch 3500 fish per year.

(a)   Using these assumptions derive a recurrence relation to model the population.

(b)   Without solving the recurrence relation describe the long-term behaviour of the population in the two cases when the initial population is 5000 or 15000.

(c)   Use a calculator or computer and software package (such as a spreadsheet) to solve the recurrence relation to verify your results in (b).

3.   A colony of birds currently has a stable population. Prior to this situation the population increased from an initially low level. When the population was 10000 the proportionate birth rate was 50% per year and the proportionate death rate was 10% per year. When the population was 20000 the proportionate birth rate was 30% and the proportionate death rate was 20%.

A model of the population is based on the following assumptions:

(i)   There is no migration and no exploitation (such as shooting).

(ii)   The proportionate birth rate is a decreasing linear function of population.

(iii)   The proportionate death rate is an increasing linear function of population.

Show that a model based on these assumptions and the above data predicts that the population grows according to the logistic model and find the stable population size.

Shooting of the birds is now allowed at a rate of 20% of the population per year. Derive the recurrence relation which models the exploited population. Find the equilibrium population. By solving the recurrence relation using a calculator or computer software package (such as a spreadsheet), predict how many years of hunting will take place before the population is first within 1% of the equilibrium population.

4.   The population of fish in a large lake has been stable for some time. Prior to this situation the population was decreasing from an initially relatively high level. When the population was 4000 the proportionate birth rate was 10% and the proportionate death rate was 70%. When the population was 3000 the proportionate birth rate was 30% and the proportionate death rate was 60%.

A model of the population is based on the following assumptions:

(i)   there is no exploitation and no restocking;

(ii)   the proportionate birth rate is a decreasing linear function of the population;

(iii)   the proportionate death rate is an increasing linear function of population.

Show that the model based on these assumptions and the above data predicts that the population falls according to the logistic model; find the equilibrium population size.

Restocking of the lake now takes place at a rate of 20% of the population per year. Derive the recurrence relation which models this situation and find the equilibrium population. How many years will it be before the population is within 5% of the new equilibrium population?

Sketch the graph of the population as a function of time.

3 •Mathematical Modelling in Action

3.1 Introduction

This chapter contains reprints of a selection of articles on mathematical modelling which have been published in books or journals. Each article deals with a different situation and they all demonstrate mathematical modelling in action. The author of each article has developed one or more mathematical models to describe a particular situation and has used the models to answer questions about the situation.

There are several reasons why we should study published articles on modelling. First it gives us information about, and insights into the particular situation described, and we learn something about the world which we did not know before. Secondly it shows us how other people have used mathematics to model situations and to solve problems, and we learn something about the mathematical modelling process just by observing what others have done. Thirdly we may learn some new mathematics or see familiar mathematics being used in a new situation. Finally, and perhaps most importantly, we learn how to read mathematical modelling articles *with a critical eye*, we learn to ask pertinent questions about the author's models - What has (s)he done? Why did (s)he do this? How did (s)he do it? Is it valid? -and we learn not to take everything on trust, that sometimes people make mistakes and get things wrong and that it could be important to us to find these mistakes and to correct them.

When confronted with a new problem, the first thing a professional mathematician does is to see if someone else has already solved the problem, or one like it. (S)he does this, in the first instance, by reading books and journals. So *reading* mathematics is part of the way of life of a mathematician. And this is more than just a "skim read". It is a careful, critical read with the reader working through much of the mathematics for themselves. In this way (s)he comes to a fuller understanding of what has been written and how the author has solved the problem. It is also important, in the everyday world, that politicians and journalists should not be able to "blind us with science". In these circumstances, having the ability to read with a critical eye and to ask pertinent questions will save us from being hoodwinked!

3.2 How to use the articles

In the first case study we shall give you an article to read and then we shall take you through it slowly, asking pertinent questions as we go along and pointing out important features of the article. Hopefully this will give you an idea of how *we* read the article and came to an understanding of it.

Each of the remaining case studies contain an article for you to read, together with a set of questions which we consider pertinent and which should tease out your understanding of the article. You should attempt to understand each article before looking at our questions, otherwise you will read the article only with a view to answering the questions and you may miss other things in the article. As you read it, you should jot down your own pertinent questions and your answers to these questions. In this way you will probably anticipate most of our questions and will probably ask many more besides. It would also be useful for a group of students to pool their ideas and questions and to discuss these with one another before looking at the set questions. You can usually learn quite a lot from the other people in your group. But do this *yourselves*; do not depend on your teacher for either the questions or the answers.

As you read the articles keep the following questions in mind:-

- What modelling assumptions has the author made?
- Has (s)he stated these explicitly or are they implied by something (s)he has written?
- Are they *reasonable* assumptions? Would you be happy to make these assumptions or would you make others?
- Does the author justify the assumptions? If not can you give a justification of them?
- Can you follow the mathematics through from line to line? Has (s)he made any mistakes?
- Are the numerical calculations correct?
- Can you explain all statements like "it can be shown that ..." or " it follows from the above that ..."?
- Has (s)he attempted to validate the model by comparing with observations? If not, can you suggest an experiment that could be carried out?
- Does (s)he draw valid inferences and conclusions from the work?
- Is sufficient background information to the problem given in the article or is it necessary to find out things from other sources?

There is quite a lot here to think about, but don't worry, most of it is just natural curiosity and you will get better at asking the right questions after you have studied a few articles.

Now read through the article in Case Study 1 and study our commentary on it. The numbers in the left margin are line numbers which are referred to in the commentary.

CASE STUDY 1 The Percy-Grainger Feat

This article was inspired by Chapter 8 of Modelling with Projectiles by D Hart and T Croft.

The Percy Grainger Feat

1. Introduction

Percy Grainger was born in 1882 and was one of Australia's best known composers. His hobby was sport – he loved running – and he was widely known as the jogging pianist! The story goes that he was able to throw a cricket ball over the roof of his

5   house and to run through the house and catch the ball on its way down. This article attempts to model this feat so that we can decide whether or not it really is feasible.

2. The Model and a Solution

Fig 3.1 illustrates the situation.

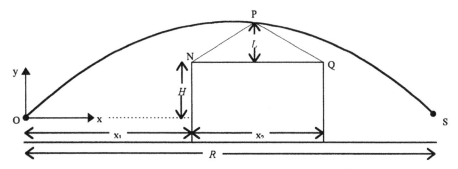

Fig 3.1 The Percy Grainger Feat

The origin 0 of the co-ordinate system is at the point of projection of the ball. x_1 is
10   the distance of the thrower from the house and x_2 is the distance through the house. H is the vertical distance from the x-axis to the top of the wall and L is the further distance to the apex of the house. R is the range of the ball assuming that it is caught at the same height from which it was thrown.

We assume that the ball is thrown with speed V_0 at an angle of projection θ to the
15   horizontal. Typical values for these constants are $H = 5$, $L = 2$, $x_1 = 10$, $x_2 = 10$ and $V_0 = 20$ in appropriate units.

From standard projectile formulae, we known that, in the absence of air resistance, the time of flight is $2V_0\sin\theta/g$ and the horizontal range is $2V_0^2 \sin 2\theta / g$. The equation of the flight path is

$$y = x\tan\theta - \frac{gx^2}{2V_0^2}(1 + \tan^2 \theta)$$

20   Assume that the thrower is stationary when the ball is released.

What angle should the ball be thrown at?

There are several things to bear in mind. If θ is too small, the ball will hit the side of the house or the near-side of the roof. If θ is too large, the ball may clear the apex of the roof but hit the far side of the roof on the way down. If the thrower runs too
25   slowly, he may not arrive in time to catch the ball, although we are not too worried if he arrives too early and has to wait for the ball to come down.

Now, the horizontal speed of the ball is constant at $V_0\cos\theta$ so clearly this must be the minimum average speed of the thrower to allow him to arrive at the point S in time to catch the ball.

30

In addition there are other constraints to be met:

1. when $x = x_1$, $y > H$;

2. when $x = x_1 + x_2/2$, $y > L + H$;

3. when $x = x_1 + x_2$, $y > H$.

Imposing constraint (2) yields the quadratic inequality

35

$$2.8125T^2 - 15T + 9.8125 < 0$$

where $T = \tan\theta$ and the typical values of the parameters given above have been used.

The solution set of this inequality is

$$0.763 < T < 4.57$$

which means that

40

$$37.34° < \theta < 77.66°$$

The reader can verify that constraint (1) is automatically satisfied when θ lies in the range.

Imposing constraint (3) implies that

$$T^2 + 4T + 2 < 0$$

45

which leads to

$$30.36° < \theta < 73.68°$$

Hence the set of possible angles of projection which satisfies the three constraints is

$$37.34° < \theta < 73.68°$$

50

Now the thrower must run at an average speed greater than $V_0\cos\theta$. Since $\cos\theta$ is a decreasing function of θ in the range $0° < \theta < 90°$, then we should take θ as large as possible, i.e. $\theta = 73.68°$. Then the speed required of the runner is $20\cos\theta$ or 5.62ms^{-1}. When $\theta = 45°$, the average running speed is 14.14ms^{-1}, but speeds like this have not yet been achieved! So, while it is easier to throw the ball at $45°$ and still clear the roof, there is no possibility of catching it as it descends.

3. The Bounding Parabola

When a projectile is launched from an origin 0 with speed V_0 there are certain regions in the xy plane which are within the range of the projectile and other regions which are out of range. In the absence of air resistance, the curve in the plane which separates these regions is known as the bounding parabola or the enveloping parabola or the parabola of safety. Let us try to find the equation of this curve.

The equation of the trajectory of a projectile is

$$y = x\tan\theta - \frac{gx^2}{2V_0^2}(1 + \tan^2\theta)$$

which can be written in the form

$$x\tan\theta - \frac{gx^2}{2V_0^2}(1 + \tan^2\theta) - y = 0 \ .$$

A point (X, Y) is within range if a value of θ can be found such that X and Y satisfy this equation

$$X\tan\theta - \frac{gX^2}{2V_0^2}(1 + \tan^2\theta) - Y = 0 \ .$$

This is a quadratic equation for $\tan\theta$ and can be written in the form

$$\tan^2\theta - \frac{2V_0^2}{gX}\tan\theta + \left(1 + \frac{2YV_0^2}{gX^2}\right) = 0 \ .$$

A quadratic equation like this may have real roots, equal roots or complex roots. The significance of complex roots is that there is no angle θ which will give a trajectory which passes through (X, Y). If the roots are real and unequal, then there are two trajectories, with different angles θ which will pass through (X, Y). If the roots are equal then there is only one trajectory through (X, Y). This case separates the accessible region from the inaccessible region. Applying the condition for equal roots, we obtain

$$\frac{4a^2}{X^2} = 4\left(1 + \frac{2Ya}{X^2}\right)$$

where $a = V_0^2 / g$.

Simplifying we get $a^2 = X^2 + 2aY$ and conclude that the equation of the bounding parabola is $2ay = a^2 - x^2$.

Commentary

The first question that occurred to us was "Is this feat possible at all?" *Is* it possible to throw a cricket ball over a house and run through to catch it? Is it possible to throw a ball that high and that distance, and is it possible to run that distance in the time available?

What do we know about throwing a cricket ball, and what do we know about sprinting? Well, professional cricketers can readily throw a cricket ball from the boundary of a cricket pitch to the wicket at the centre, a distance of about 75 yards, or about 70m. From this we can estimate the speed at which it is possible to throw a cricket ball. Also top class sprinters can run 100m in about 10 seconds. Lesser mortals will take a few seconds longer. This should help us estimate the sprinting speed of Percy Grainer and we can work out whether it is fast enough or not.

But what about the modelling? Clearly the author is using a projectile model. Since "standard projectile formulae" is referred to in the line labelled [17]. You probably know them as

$$x = V_0 t \cos\theta \qquad\qquad\qquad (1)$$

$$y = V_0 t \sin\theta - \frac{1}{2} gt^2 \qquad\qquad\qquad (2)$$

You can work out the time of flight by putting $y = 0$ in (2) and solving for t. Then R can be obtained by substituting the positive value for t into (1).

What are the modelling assumptions implicit in using a projectile model? It is important to consider these so that we can decide if they are reasonable assumptions to make in this situation.

The projectile formulae are obtained by solving the differential equations

$$m\,\frac{d^2x}{dt^2} \;=\; 0, \quad m\,\frac{d^2y}{dt^2} \;=\; -mg \qquad\qquad\qquad (3)$$

subject to the initial conditions

$$x = y = 0, \quad \frac{dx}{dt} = V_0 \cos\theta, \quad \frac{dy}{dt} = V_0 \sin\theta.$$

Here we are using Newton's 2nd law of motion

$$\text{mass} \times \text{acceleration} = \text{force}$$

and we are assuming

(i)   that the projectile (the ball) is a particle and so does not have a moment of inertia
and does not spin;

(ii)   that the coordinate system we are working in is not accelerating and so we are
ignoring the fact that the earth is spinning on its axis;

(iii)   that Newton's 2nd law is a valid law, that is, we can explain acceleration by
considering the forces acting.

We are further assuming in equations (3)

(iv)   that the only force acting is gravity, and so we are ignoring air resistance;

(v)   that the force of gravity is constant both in magnitude and direction, and so we are
taking an approximation to Newton's universal law of gravity.

Assumption (v) is an interesting one. Newton's universal law of gravity says that the force of
gravity between two bodies is directed along the line between their centres of mass and is
proportional to $1/r^2$ where r is the distance between the centres of mass. In assumption (v) we
are effectively saying that for projectile motion the earth is flat since the direction of the force
(which is perpendicular to the earth's surface) changes by less than $50/r$ radians, which is very
small since $r = 6.4 \times 10^6$m. Also, the ratio of the magnitude of the force of gravity on the ball
at the top of its flight to the magnitude at the bottom is approximately

$$\left(\frac{r}{r+30}\right)^2$$

which again is very small.

So clearly assumption (v) is a reasonable one. So also is assumption (iii) because we know
that Newton's 2nd law is very good for modelling the motion of small bodies. (In fact it wasn't
until astronomers in this century noticed small discrepancies in the predicted orbit of the
planet Mercury that mathematicians found it necessary to develop a new model. This
eventually appeared as Einstein's General Theory of Relativity).

We can accept assumptions (i), (ii) and (iv) for now but these should be investigated later
when you have studied moments of inertia, rotating co-ordinate systems and air resistance.
These are all small effects as far as this problem is concerned and so it is reasonable to ignore
them as our author has done.

Next, have you worked out the significance of the three inequality constraints (1), (2) and (3)
given in the article at [31]-[33]? These imply that the trajectory of the ball will clear points N,
P and Q on the house respectively and so the ball will go over the house.

At [16] in the text the author refers to "appropriate units". These are probably SI units
involving metres and seconds, but don't you think the author should have been more explicit?

How does the *size* of this house compare with yours? Did the author just pick these numbers
for convenience or are they reasonable dimensions for a house of bungalow? The height of the
house is 5 + 2 + 2 (the height of the thrower) which equals 9m or about 30 ft. This would be a

two storey house with a roofspace. The house is also 10m or about 33ft from front to rear. A good sized room in a house is about 12ft x 12ft and so this house is of the right size.

The suggested throwing speed is $V_0 = 20\text{ms}^{-1}$. How does this compare with the speed of throw of our professional cricketer? Suppose 70m is the maximum throw of a cricketer. This implies a speed of projection given by

$$V_0 = \sqrt{gR} = \sqrt{10 \times 70} = 26\text{ms}^{-1}$$

So 20ms^{-1} should be achievable by our hero. (His maximum range will then be $R = V_0^2/g = 40\text{m}$.) The inequality at [35] can be derived fairly easily but some care must be exercised in solving it.

Start with the equation at [19], put $g = 10$, $V_0 = 20$ and $\tan \theta = T$ to get

$$y = Tx - x^2(1 + T^2)/80.$$

When $\qquad x = 10 + 10/2 = 15$ we require $y > 2 + 5 = 7$

therefore $\qquad 15T - 225(1+T^2)/80 > 7$

or $\qquad 2.8125T^2 - 15T + 9.8125 < 0$

The graph of the left hand side is a cup shaped parabola which is negative for values of T between the roots.

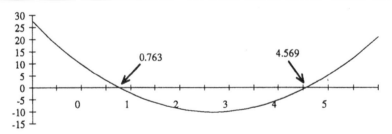

Hence [35] is satisfied when

$\qquad 0.763 < T < 4.569$

or $\qquad 37.36° < \theta < 77.66°$

Inequality (3) at [33] leads to

$\qquad 5T^2 - 40T + 25 < 0$

or $\qquad 34.33° < \theta < 82.22°$.

Since $34.33 < 37.36$ and $77.66 < 82.22$ then condition (3) is automatically satisfied when condition (1) is satisfied as the author claims.

Condition (2) leads to

$$30.36° < \theta < 73.68°$$

and combining (1) and (2) give

$$37.36° < \theta < 73.68°.$$

So Perry must throw his cricket ball at 20ms^{-1} at some angle between 37.36° and 73.68°.

The time of flight of a projectile is $2V_0 \sin \theta/g$ and this increases with θ. So it would give him more time to run through the house if he could achieve a large angle of projection. Have you tried this? It gets harder to give the ball the necessary initial speed as θ increases.

However assuming that V_0 is always 20, the time of flight lies between 2.43 seconds (at $\theta = 37.36°$) and 3.84 seconds (at $\theta = 73.68°$).

The average sprinting speed of the thrower must be greater than the horizontal speed ($V_0 \cos\theta$) of the ball. Since $37.36° < \theta < 73.68°$ then the minimum sprinting speed lies in the range 5.62ms^{-1} to 15.90ms^{-1}.

The distance to be covered lies between 21.6m (at $\theta = 73.67°$) and 40m (at $\theta = 45°$). (Did you spot this?)

If the ball is thrown at 73.67° then the thrower must cover 21.6m in at most 3.84 seconds i.e. he must run at average speed of at least 5.62ms^{-1}.

From a standing start, the first part of a short sprint is run at an average speed slower than the average speed for the whole sprint. Thus if a sprinter can cover 100m in, say 12 seconds, (an average speed of 8.3ms^{-1}) then he might only achieve an average speed of 7.0ms^{-1} over the first 25m. He would then be required to throw the ball at 20ms^{-1} at an angle of at least 70° in order to achieve the feat.

Exercise 1

1. The angle of 73° obtained above is rather too large for practical purposes; it would be difficult to achieve a reasonable projection speed at such an angle. Given that the average running speed is 10ms^{-1}, what angle should the ball be thrown at? Is a practical solution feasible?

2. Given that the size of the house is fixed, and the projection speed V_0 is 20ms^{-1}, what variables can be changed? (For example, you could throw from further away.) Estimate some typical parameters, e.g. the size of your house, and repeat the above analysis.

3. For the same house, suppose, now, that V_0 can vary up to a maximum of 30ms^{-1}. Investigate the effects of variation of speed and distance from the house when the average running speed is no greater than 10ms^{-1}. Remember the ball must clear the house and you must be able to run fast enough to catch it.

4.   What is the size of the largest house for which the problem has a solution? let the throwing speed not exceed $30ms^{-1}$ and the running speed not exceed $10ms^{-1}$.

Have we answered all the pertinent questions?

>   What assumptions have been made?
>   Are they justified or justifiable?
>   Is the mathematics correct?
>   Has the model been validated or could it be validated?

Yes we have, and we now have some understanding of a particular problem. To investigate it we have had to make use of our own previous knowledge or knowledge not included in the article. Nevertheless it was not too difficult to find out about cricketers and sprinters and the background information relating to projectiles can be used time and time again.

CASE STUDY 2   Safe Driving Speeds on Newly Surfaced Roads

There follows two extracts from Modelling with Projectiles. The extract from Chapter 3 introduces "The enveloping parabola" and that from Chapter 9 makes use of this to model the "Safe driving speeds" problem.

Extract from: Modelling with Projectiles by Derek Hart and Tony Croft
Chapter 3. Ellis Horwood, 1988.

The Enveloping Parabola

3.1   Derivation Of The Enveloping Parabola

It is intuitively obvious that if a projectile is launched from an origin O with speed V_0 there are certain regions of the xy plane within its range, and there are other regions which are not. In the absence of air resistance the dividing curve between these regions has become variously known as the Bounding Parabola, the Enveloping Parabola, or the Parabola of Safety. In this chapter we develop the theory of this special curve and show how it can be used to solve many maximum-range-type problems.

Consider the family of curves obtained when we plot the trajectories of a projectile, launched with a given, fixed speed V_0, at various angles of projection θ. A number of such curves are shown in Fig 3.2.

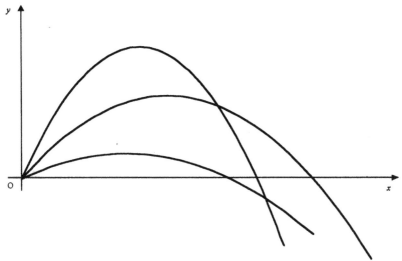

Fig 3.2 – Members of a family of curves.

If we add more and more members of the family, it soon becomes apparent that there is a curve, which is not a trajectory, which touches all the trajectories. This is the dashed curve in Fig 3.3.

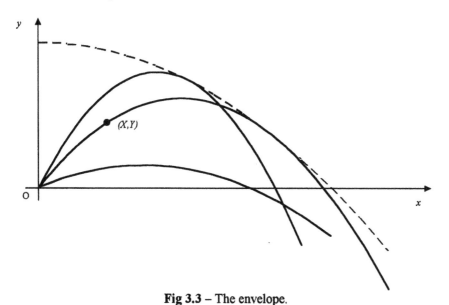

Fig 3.3 – The envelope.

Since all the trajectories lie under this curve, it is the dividing curve between so-called 'accessible' regions and 'inaccessible' regions of the xy plane. We now develop its equation in two ways.

Recall that the equation of the trajectory is given by

$$y = x\tan\theta - \frac{gx^2(1+\tan^2\theta)}{2V_0^2} \tag{3.1}$$

which we can write in the form

$$x\tan\theta - \frac{gx^2(1+\tan^2\theta)}{2V_0^2} - y = 0 \tag{3.2}$$

A point (X, Y) is accessible if some trajectory passes through it as shown in Fig 3.3. In this case the corresponding angle of projection is found by solving

$$X\tan\theta - \frac{gX^2(1+\tan^2\theta)}{2V_0^2} - Y = 0 \tag{3.3}$$

This is a quadratic equation for $\tan\theta$ which can be written in the form

$$\tan^2\theta - \frac{2V_0^2\tan\theta}{gX} + \left(1 + \frac{2YV_0^2}{gX^2}\right) = 0 \tag{3.4}$$

It is well known that a quadratic equation $ax^2 + bx + c = 0$ possesses

$$\text{real roots if } b^2 - 4ac > 0$$
$$\text{equal roots if } b^2 - 4ac = 0$$

and

$$\text{complex roots if } b^2 - 4ac < 0$$

In the last case there are clearly no real values of $\tan\theta$ for which a trajectory passes through (X, Y), and thus (X, Y) is in the inaccessible region of the xy plane. On the other hand, if there are two real distinct roots then (X, Y) can be reached along two different trajectories. If only equal roots exist then (X, Y) is reached by only one trajectory. This case divides the inaccessible from the accessible regions, so it is precisely this case which represents the enveloping parabola. Applying the condition for equal roots we find

$$\frac{4a^2}{X^2} = 4\left(1 + \frac{2Ya}{X^2}\right)$$

where $a = V_0^2/g$. This is rearranged to

$$a^2 = X^2 + 2Ya$$

We conclude that the equation of the enveloping parabola is

$$2ay = a^2 - x^2 \tag{3.5}$$

Extract from: Modelling with Projectiles by Derek Hart and Tony Croft
Chapter 9. Ellis Horwood, 1988.

Safe Driving Speeds on Newly Surfaced Roads

9.1   The Problem

Cheshire County Council in common with many local authorities recommends that
drivers limit their speeds to 20 mph when travelling over roads which have been
newly dressed with a bituminous binder and stone chippings. One unavoidable snag
with this method of maintenance is that some surplus stones remain on the road until
they are swept up some hours after laying. The advisory limit is intended to avoid the
situation where stones are caused to fly into the path of other vehicles and so cause
damage to paintwork and even breakage of windscreens. In this chapter we present
various mathematical models which describe some aspects of this situation, and we
argue that a 'safe' speed is not significantly lower than that recommended.

9.2   Assumptions

1. The wheel does not skid. Owing to a variety of factors such as the nature of the
   tyre, the tackiness of the binder, the weight and size of the vehicle, stones are
   thrown up at random speeds not greater than the speed of the vehicle. We therefore
   assume the simplest case in which stones are projected with the speed of
   the vehicle.
2. In practice it is observed that the direction of projection is random, but in the first
   simple model we assume that a stone remains in the vertical plane containing the
   direction of motion of the vehicle.
3. Vehicles travel at constant speed, their separation from the vehicle in front being
   not less than that recommended by the Highway Code.
4. Air resistance is negligible.
5. Drivers are not unduly worried by stones hitting the front bumper, grille, etc.

In the solution we shall make use of the bounding parabola.

9.3   Solution

A stone is projected from the origin 0 at time $t=0$ and moves along a typical
trajectory as shown in Fig 3.4. The front of the bonnet of the following car, FC, has

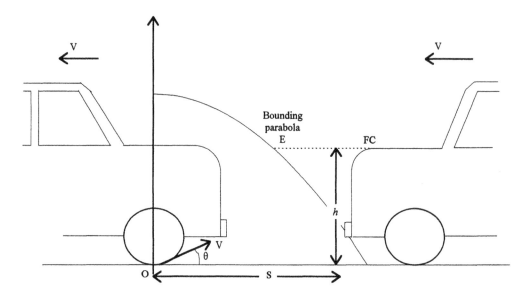

Fig 3.4

coordinates (S,h). As a first attempt we will argue as follows.

There is certainly a danger of the bonnet, windscreen or top of the car being hit by the stone if FC has crossed the bounding parabola, i.e. if FC is within the bounding parabola at time t_E, the time of flight of the stone from O to E.

After time t_E the position of FC is

$$S - Vt_E$$

and so the vehicle is in danger of being hit if

$$S - Vt_E < X_E \tag{9.1}$$

where X_E is the x coordinate of E.

9.4   Separation Distances

An expression for S is now required and experience shows this to be related to the speed of travel V. We will assume that drivers adhere to the Highway Code recommended separation distances. This assumption enables us to determine S as a function of V in the following way.

The Highway Code provides the information depicted in Fig 3.5. The data show that S can be divided into two parts, namely (a) the thinking distance S_T (which is the distance travelled in the time it takes for the driver to react and depress the brake pedal), and (b) the braking distance S_B (which is the distance travelled whilst braking from speed V to rest).

Shortest stopping distances

At 30 m.p.h.

Thinking distance	Braking distance	Overall stopping distance
30 ft.	45 ft.	75 ft.

At 50 m.p.h .

Thinking distance	Braking distance	Overall stopping distance
50 ft.	125 ft.	175 ft.

At 70 m.p.h.

Thinking distance	Braking distance	Overall stopping distance
70 ft.	245 ft.	315 ft.

Fig 3.5 – Shortest stopping distances.

It is clear from the data that $S_T = V$ feet.

To obtain the formula between S_B and V it is useful to assume a relationship of the form $S_B = kV^\alpha$ where k and α are constants. Then $\log S_B = \alpha \log V + \log k$.

Plotting $\log S_B$ against $\log V$ should yield a straight-line graph if the assumed relationship is true. From such a graph the values of α and k can be determined. This graph is illustrated in Fig 3.6.

The graph is indeed a straight line of gradient 2, so that the assumed relationship between S_B and V is true, and the value of α is 2. Furthermore the value of k can be shown to be 0.05. We conclude that $S = V + 0.05V^2$, where S is measured in feet and V in mph. However, our model requires the use of SI units. It is straightforward to show that the equivalent formula when S is measured in metres and V in metres per second is

$$S = 0.682 + 0.076V^2 \qquad (9.2)$$

9.5   Solution Continued

Since E lies on the bounding parabola its x coordinate is obtained from (3.5) as

$$X_E = \sqrt{a(a - 2h))} \qquad (9.3)$$

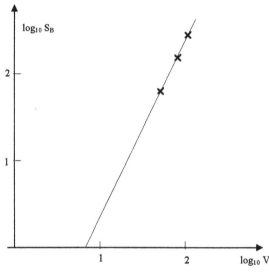

Fig 3.6

where $a = V^2/g$. Furthermore from (3.8)

$$\tan\theta = a/x \qquad (9.4)$$

whence $\cos\theta = x / \sqrt{(a^2 + x^2)}$ and since $x = Vt\cos\theta$ we obtain

$$t_E = \frac{\sqrt{(a^2 + X_E^2)}}{V} \qquad (9.5)$$

Combining (9.1), (9.3) and (9.5), we have the result that the following vehicle is in danger of being hit by the stone if

$$S - \sqrt{(2a(a - h))} < \sqrt{(a(a - 2h))} \qquad (9.6)$$

Therefore from (9.2), there is danger if

$$\sqrt{g(0.682 + 0.076V)} < \sqrt{(V^2/g - 2h)} + \sqrt{\left(2\left(\frac{V^2}{g} - h\right)\right)} \qquad (9.7)$$

9.6 Interpretation of the Result

If the equality given by (9.7) is solved numerically for $h = 0.75$ m (typical of many popular makes of car) we find that if a driver exceeds 5.531 ms^{-1} (12.372 mph) he is in danger of being hit. It might be argued that even though the car remains outside the bounding parabola at time t_E, at some later time when it is within the bounding parabola it could still be hit by a stone which has taken time $t_h > t_E$ to reach height h. It can easily be shown that the time taken for a stone to rise and then fall to height h is

$$t_h = \frac{V\sin\theta + \sqrt{(V^2\sin^2\theta - 2gh)}}{g}$$

(9.8)

from which we deduce that t_h is an increasing function of θ. Taking $V = 5.531$ ms^{-1} and $h = 0.75$ m, equations (9.3) and (9.4) imply that θ_E, the angle of projection which takes a stone to the point E on the bounding parabola, is about 54°. Consequently a greater time to reach a height h must arise from any angle of projection greater than this. However, observation shows that the wheel arches and rubber flaps of the vehicle constrain angles to be much less than 54°. We therefore conclude that if vehicles travel at 5.531 ms^{-1} with a separation of at least that recommended by the Highway Code, there is no danger of being hit. The value of 5.531 ms^{-1} obtained in this way is rather lower than the advisory limit recommended by local authorities, and in practice, drivers are reluctant to slow down even to 20 mph and would almost certainly ignore requests to slow down even further.

9.7 Model 2

Suppose, now, that we no longer assume a Highway Code separation. Consider a stone projected from O (Fig 3.4). The time it takes to reach height h as shown is given by (9.8). Its x coordinate is then

$$x_h = V\cos\theta t_h$$

(9.9)

If the following car has separation S_1 given by

$$S_1 = x_h + Vt_h$$

(9.10)

then a stone projected at angle θ will reach the bonnet at time t_h. Our procedure is as follows: choose V and calculate the smallest angle θ_{MIN} that can cause a stone to reach height h at this speed, then gradually increase θ from this smallest value up to $\pi/2$, at each stage calculating S_1 from (9.10). Simultaneously we calculate $S = 0.682V + 0.076V^2$, the Highway Code separation. We then gradually increase V and repeat the whole procedure. It soon becomes apparent that at very low speeds, $S_1 < S$ for all angles of projection. This means that if a vehicle travels at the recommended separation it must be outside the range of the furthest stone to reach height h. Eventually, as V is increased there comes a point when $S_1 > S$ for some angles of projection, meaning that a vehicle travelling at the recommended separation can be hit on the front of the bonnet by a stone at some angle. This searching problem is easily handled on a microcomputer. For $h = 0.75$ m it reveals that if V is less than about 5.17 ms^{-1}, a car travelling at the recommended separation is safe.

Above this speed there is the possibility of a stone hitting the car. Again a value of 5.17 ms^{-1} is unrealistic in practice, but we can argue that instead of allowing θ to vary from θ_{MIN} to $\pi/2$, we constrain it to lie between θ_{MIN} and $35°$, say. The searching routine so modified yields a speed of approximately 8 ms^{-1} or 18 mph. Measurement of the overhang of the boot on many makes of car leads us to the conclusion that $35°$ is a maximum value, stones projected at a higher angle hitting the underside of the vehicle. Many saloon-type cars have a much smaller maximum. (We are of course neglecting tractors, etc.) This second model leads us to the conclusion that for most types of vehicle travelling at the Highway Code recommended separation, a safe driving speed is not significantly lower than that recommended by the local authority.

Questions

(i)   By applying the condition for equation (3.4) to have equal roots, show that the equation $a^2 = X^2 + 2Ya$ is correct.

(ii)   Using the data in Fig 3.5, calculate log S_B and log V, and plot the data on a graph, the first quadrant of which is sketched in Fig 3.6.

Use your graph to find α and k. Show your working clearly and verify that the results in the article are correct.

(iii)   Find the printer's mistake in equation (9.2).

(iv)   Showing your working clearly, verify that the values 0.682 and 0.076 in equation (9.2) are correct.

Note: 1 mile = 1609 metres; 1 metre = 3.282 feet.

(v)   Explain why the equation

$$S_B = 0.076V^2$$

where S_B is in metres and V is in metres per second implies that the car is undergoing a constant deceleration.

Calculate the value of this deceleration.

If the mass of the car is 100kg, calculate the braking force on the car.

(vi)   Solve equation (3.4) for tan θ in the "equal roots" case and hence obtain equation (9.4).

(vii)   Derive equation (9.8).

(viii) Find a formula for θ_{MIN} (which is referred to in the article 2 lines after equation (9.10)).

(ix)   Why do drivers following a tractor need to be careful?

        What advice would you give them?

(x)   Show that a car following a tractor, both travelling at speed V, will certainly be safe from flying stones if the distance between them is $4V^2/g$.

(xi)   Modify the model to meet the needs of a driver who is worried by stones hitting the front bumper or grill (Assumption 5).

Note: Questions (i) to (x) are taken from the NISEAC A level paper, Further Mathematics (Mode 2) Paper 3, 1990 and are reproduced with permission.

CASE STUDY 3 Drug Therapy

Extract from: Applying Mathematics by D N Burghes, I Huntley and J MacDonald. Chapter 18. Ellis Horwood, 1982.

Note: tables and figures are excluded from the line numbering system.

18   Drug Therapy

18.1 Introduction

Patients with asthma have constriction of the airways in the lungs and consequent difficulty in breathing out. This ailment can be alleviated by introducing the drug theophylline into the bloodstream. This is done by injecting another drug, aminophylline, which the body quickly converts to theophylline. Once present in the blood, however, the drug is steadily excreted from the body via the kidneys. In other words, the system 'leaks', and unless there is replenishment the quantity of drug in the blood will fall.

10   It is known from experiments that theophylline has hardly any therapeutic effect if its concentration in the bloodstream is below 5 mg/l, and that concentrations of above 20 mg/l are likely to be toxic. The problem is to administer the drug in such a way that the concentration remains within the therapeutic range between 5 mg/l and 20 mg/l.

Experimental measurements can be obtained by injecting a known dose of theophylline into a patient, allowing time for it to diffuse throughout the bloodstream, and finding the drug concentration in blood samples taken at regular intervals. One then knows the initial quantity put into the bloodstream and the concentrations at various later times, but there is no time for which both the quantity and the concentration are known. The following data come from such an experiment.

Initial quantity = 300 mg

Concentration, mg/l	Time, hours
10.0	1
7.0	3
5.0	5
3.5	7
2.5	9
2.0	11
1.5	13
1.0	15
0.7	17
0.5	19

Another experimental observation is that the apparent volume of distribution (V litres) and the patient's weight (W kg) are connected by the simple relationship $V = \frac{1}{2}W$. Thus the dose necessary to achieve a required initial theophylline concentration can be inferred from the patient's weight alone: the dose D mg to obtain a 12 mg/l concentration in a 50 kg patient is obtained from

$$D/V = 12 ,$$

so that $D = 12\ V = 6\ W = 300$ mg.

Assuming that the rate at which the drug is removed by the kidneys is proportional to the amount of drug in the body, obtain a differential equation which describes the concentration $c(t)$ at any time t. Solve this equation, and apply the answer to the data given above to see how well the model agrees with reality.

It is clear that in order that the drug concentration remains inside the therapeutic range a series of injections must be given, and the desired concentration pattern may be as in Fig 3.7.

By considering a very simple case where equal sized doses D mg are administered at equal intervals of time T hours, confirm that it is possible to achieve a saturation as in the Fig 3.7. Is it possible to choose the dose D and the interval T to remain in the therapeutic range? What would you advise for the patient mentioned above? Is there any advantage in giving a larger dose initially and then a series of doses of size D as above?

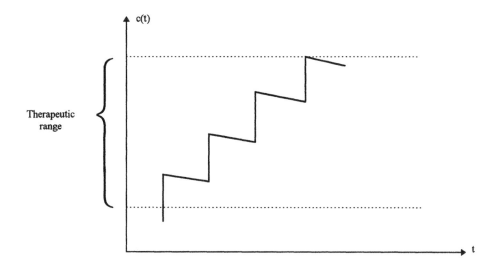

Fig 3.7

18.2 Possible solution

When given a question involving data, it is always a good idea to plot them.
Doing so in this case gives the graph shown in Fig 3.8. This reminds us of an
exponential curve, and replotting as ln c against t (Fig 3.9) confirms this guess.

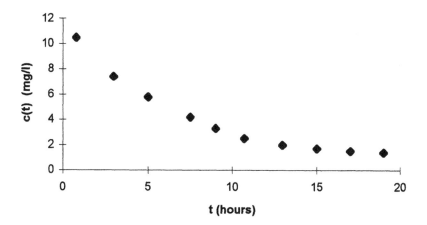

Fig 3.8

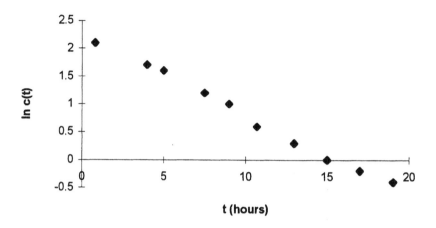

Fig 3.9

Thus we expect that our theory will yield something like $c(t) = c_0\, e^{-kt}$. To confirm this result theoretically we note that the drug is removed by the kidneys at a rate proportional to the amount present. Thus if $y(t)$ is the amount of drug present at time t, then

50 $\qquad dy/dt = -ky$.

We also know, though, that the amount of drug and the concentration are connected via the apparent volume:

$\qquad c(t) = y(t)/V$.

Thus $dc/dt = -kc$, giving us

$\qquad c(t) = c_0 e^{-kt}$ as anticipated.

To complete this part of the analysis, we must evaluate the parameters c_0 and k. The linear graph, Fig 3.9, is $\ln c$ against t. However, we have

$\qquad \ln c = \ln c_0 - kt$,

so the intercept on the graph is $\ln c_0$ and the slope is $-k$. This gives

60 $\qquad c_0 = 12$ mg/l, $\quad k = 0.17$/hour

and agrees well with the data given for the 50 kg patient.

We now assume that another dose of D mg (where $D/V = c_0$) is given at time T. This will yield the form of concentration with time shown in Fig 3.10,

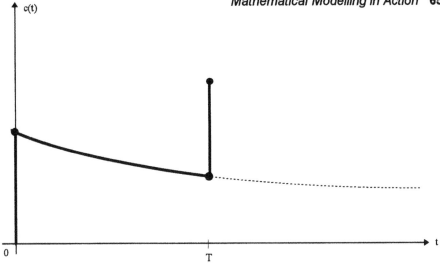

Fig 3.10

where we assume that the drug is instantaneously dispersed throughout the body. We know that $c(t) = c_0 e^{-kt}$, so we have

$$c(T) = c_0 e^{-kT} \quad \text{and} \quad c(T^+) = c_0 + c_0 e^{-kT}.$$

This new amount of drug now decays exponentially as before, so that

$$c(t) = c_0(1 + e^{-kT}) e^{-k(t-T)} \quad \text{for} \quad t > T.$$

Thus

$$c(2T) = c_0(1 + e^{-kT}) e^{-kT}$$

$$= c_0(e^{-kT} + e^{-2kT})$$

and $c(2T^+) = c_0(1 + e^{-kT} + e^{-2kT})$ after another dose. It is now clear that we can generalise this result to any number of doses. We obtain

$$c(nT^+) = c_0(1 + e^{-kT}) + \underline{\quad} + e^{-nkT})$$

which tends to the value $c_0/(1 - e^{-kT})$ for large n.

This has all become rather mathematical, so let's bring it back to earth by trying to decide the dose D and interval T to keep the concentration in the therapeutic range $5 \le c(t) \le 20$. For our 50 kg patient we already have $k = 0.17$/hour, and it seems fairly clear that we want the limit $c_0/(1 - e^{-kT})$ above to be 20 mg/l. Thus $c_0 = 20(1 - e^{-kT})$, and so we can choose convenient values for T and obtain the corresponding dose $D = 25 \, c_0$. This gives the following table, where we have also included the concentration just before the second dose.

T hours	*D* mg	*c(T)* mg/l
1	78	2.6
2	144	4.1
3	200	4.8
4	247	5.0
6	320	4.6
8	372	3.8
12	435	2.3
16	467	1.2
18	477	0.9
24	492	0.3

[Note that we could also have graphed *D* against *T*, but in some ways the table above is more useful. This is because the hospital administering the treatment will want a regular timetable for giving the injection, probably every 4, 6, or 8 hours, and so will not be interested in other values.]

We now see that if we choose *T* = 4 hours the concentration *never* leaves the therapeutic range. Thus, unless a concentration just less than 20 mg/l is preferable to one just greater than 5 mg/l, there is no point at all in giving a large dose and smaller repeat doses. Our advice to the doctor must be

"Give a dose of 250 mg every 4 hours".

Questions

(i) Use the method of separation of variables, or otherwise to solve the differential equation:

$$\frac{dc}{dt} = -kc \qquad (k \text{ constant})$$

(ii) The authors claim in line [61] that the relationship

$$c(t) = 12e^{-0.17t} \text{ mg/l}$$

"agrees well with the data given for the 50kg patient".

Verify this claim by copying and completing the following extended version of the table below line [20].

Measured concentration mg/l	Time (*t*) hours	$12e^{-0.17t}$ (to 1 decimal place)
10.0	1	
7.0	3	
5.0	5	
3.5	7	
2.5	9	
2.0	11	
1.5	13	
1.0	15	
0.7	17	
0.5	19	

(iii)   Given that $c(nT^{+})$ denotes $c(nT) + c_0$ for any positive integer n, state *briefly* how you would interpret $c(nT^{+})$.

(iv)   Given that $c(nT) = c_0 (e^{-kT} + ... + e^{-nkT})$ for all integers $n \geq 1$, show that $c(nT) < c((n + 1)T)$ for all integers $n \geq 1$.

Hence or otherwise deduce that $c(nT^{+}) < c((n+1)T^{+})$ for all integers $n \geq 1$.

(v)   It follows from the formulae given in (iii) and (iv) that
$c(nT^{+}) = c_0(1 + e^{-kT} + ... + e^{-nkT})$ for all integers $n \geq 1$.

Use this result to show that:

(a) $\lim_{n \to \infty} c(nT^{+}) = c_0/(1 - e^{-kT})$

(b) $\lim_{n \to \infty} c(nT) = c_0/(e^{kT} - 1)$

(vi)   Use the inequalities in (iv) and the result in (v)(a) to show that

$c_0 e^{-kT} \leq c(t) < c_0/(1 - e^{-kT})$ at any time T.

(vii)   Suppose that the advice given in [92] is followed and the 50kg patient is given an initial dose of 250mg of theophylline and a similar dose every four hours.

Show that

(a)   the initial concentration is 10 mg/l;

(b)   the minimum concentration at any subsequent time is just over 5 mg/l;

(c)   the concentration eventually settles down in the range from 10 mg/l to just over 20 mg/l.

Hence sketch a graph of $c(t)$ against t for $t \geq 0$.

(viii)   Suppose that the doctor treating the 50kg patient tries out the following alternative strategy.

An initial dose of D mg of theophylline is to be administered in order to achieve an initial concentration of 18 mg/l. Every T hours after this (when the concentration level has fallen to 10 mg/l), a smaller dose of d mg will be injected to raise the concentration level back to 18 mg/l.

(a)   Calculate D, T and d.

(b)   Hence sketch a graph of $c(t)$ against t for $t \geq 0$.

(ix)   Discuss the advantages and disadvantages of each of the strategies outlined in (vii) and (viii).

These questions are taken from the NISEAC A-level paper, Further Mathematics (Mode 2), Paper 3, 1988 and are reproduced with permission.

CASE STUDY 4 Rats

Extract from: Applying Mathematics by D N Burghes, I Huntley and J MacDonald. Chapter 19. Ellis Horwood, 1982.

19 Rats

19.1 Introduction

In an experiment by the biologist B. F. Calhoun (reported in *Scientific American*, (1962), **206** 139), a population of wild Norway rats was confined in a ¼-acre enclosure. There was an abundance of food and places to live, and predation and
5   disease were almost eliminated. Thus only the animals' behaviour with respect to one another remained as a factor that might affect the increase in their number.
It was noted that although the adult death rate remained constant at around 5% per month, the number of surviving infants fell as the population density increased. These data were presented as an effective change to the birth rate, so
10   that instead of producing an average 0.4 infants/month each female rat only produced $0.4\,K$ infants/month. The graph of this multiplicative factor K against population density is shown in Fig 3.11.
Draw a rough sketch of birth rate and death rate (rats/month) against population (rats), and use this to justify the existence of an equilibrium popula-
15   tion of about 200 rats. Approximate the graph for K above in a suitable way,

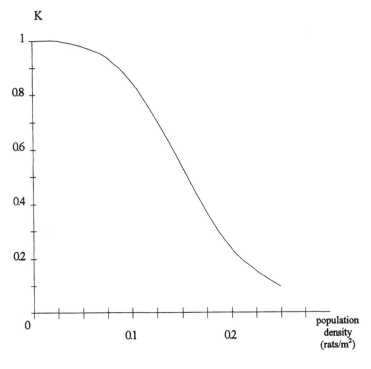

Fig 3.11

and hence derive a differential equation model. Confirm the size of the equilibrium population, and find out by how much the population has grown in one year.

19.2 Possible solution

20   In this question the information is presented in months, so we will take this as a natural timescale. We are told that the death rate is 5% per month, so we write

$$DR = 0.05\ P$$

with P being the adult rat population. To write down an expression for the birth rate, we assume – in the absence of any other information – that half the adult

25   population is female. Then the birth rate is given as

$$BR = 0.4\ K\ P/2\ .$$

To be able to sketch BR and DR against the population P we must now look at the graph for K. This is given as K (non-dimensional) against population density (rats/m^2). We also know, however, that the rats are in a ¼-acre enclosure – about

30   1000 m^2 – so we can change the axis

to the axis

and leave the curve unchanged.

We are now in a position to sketch, very roughly, the birth and death rates
against the population (Fig 3.12). [Here the birthrate curve was obtained by
35   choosing a few values of P, reading off the related value of K, and working out

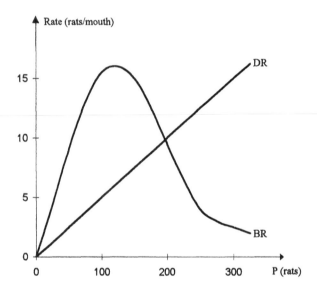

Fig 3.12

$BR = 0.2\ KP.$] We see at once that $BR = DR$ (and hence the numbers joining
and leaving the population balance, leaving the population in equilibrium) when
$P \approx 200$. To confirm this result using a differential equation model we can write

$$dP/dt = BR - DR$$

40               $$= 0.2KP - 0.05P$$

$$= 0.2P(K - 0.25).$$

Thus the population will be in equilibrium (i.e. dP/dt = 0) when K = 0.25, which can be seen to correspond to a population of about 200 rats from the original data.

45   The advantage of this latter approach is that we now have an equation which we can solve for the population P at any time t. We must firstly, though, express K in terms of P. Since we are interested mainly in the region around

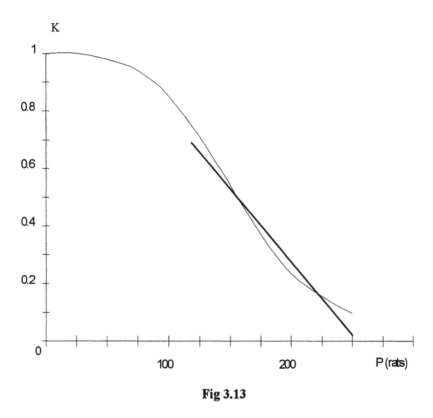

Fig 3.13

P = 200, it seems sensible to approximate K by a linear function *in this region* (Fig 3.13). We then have something like

50      $$K = 0.005\,(250 - P)$$

which can be put into our differential equation to give

$$dP/dt = 0.2P[0.005(250 - P) - 0.25]$$

$$= 0.001P(200 - P).$$

We confirm, again, the equilibrium value of 200 rats.

55   To answer the last part of the question, and find out the rat population after one year, we need to solve this equation. This is simply enough done, using separation of variables and partial fractions, and we obtain

$$\left|\frac{P}{200-P}\right| = Ae^{t/5}$$

Thus $P(t) = \dfrac{200}{1\pm\left(Ae^{t/5}\right)^{-1}}$ for $P \lessgtr 200$, and

60   writing $P(0) = P_0$ gives $A = \dfrac{P_0}{\left|200-P_0\right|}$.

After one year (i.e. 12 months) the population will be

$$P(12) = \frac{200}{1\pm\left(Ae^{12/5}\right)^{-1}}$$

which is sketched as a multiple of the initial population P_0 in Fig 3.14. The
64   graph effectively answers the original question.

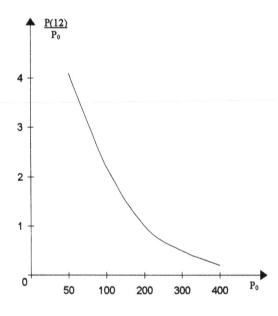

Fig 3.14

Questions

(i)   Why did the number of surviving infants fall as the population density increased?

(ii)   Given that 1 yard = 0.9144m and 1 acre = 4840 square yards, calculate the error involved in taking 1/4 acre = 1000m².

(iii)   Using the large scale version of Fig 3.13 provided below, carry out the calculations necessary to obtain the BR curve in Fig 3.12, as described in lines [34]-[36] of the article.

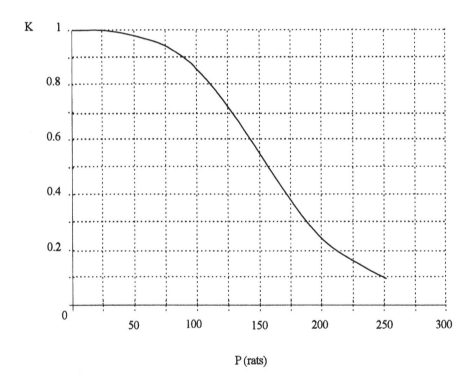

P (rats)

Hence draw your own version of Fig 3.12 and use it to find the equilibrium value of P.

(iv)   How accurate is the statement in lines [43] and [44] which starts "which can be seen ..."?

(v)   How does the author obtain the equation in line [50]? Does it relate accurately to the data? Explain your answer.

(vi)   Solve the equation in line [58] for P in each of the cases $P < 200$, $P > 200$.

(vii)   Express $P(12)$ in line [62] in terms of P_0 and hence find an expression for $P(12)/P_0$.

(viii) Show that

$$\frac{P(12)}{P_0} \cong \frac{2205}{200 + 10\,P_0}$$

and sketch the graph of $P(12)/P_0$ against P_0.

Is the y-axis $(P(12)/P_0)$ an asymptote to the graph, as could be suggested by the authors in Fig 3.14?

(ix)   In line [64] the authors claim that this graph "effectively answers the original question", which was "find out by how much the population has grown in one year".

Use your graph or the formula to answer this question when the initial population is

(a)   100,
(b)   200,
(c)   300,

and interpret your answer.

(x)   The author obtains the equation in line [62] by solving the equation in line [53]. Bearing this in mind, are the one year growth figures you have just calculated valid? Explain your answer.

(xi)   What would the graph in Fig 3.11 look like for logistic growth, modelled by the differential equation

$$\frac{dP}{dt} = kP\left(1 - \frac{P}{P_\infty}\right)$$

(xii)   Solve the differential equation in line [53] and rearrange your solution into the form of line [58]. Explain why the modulus function $|\ |$ is introduced.

Extract from: Applying Mathematics by D N Burghes, I Huntley and J MacDonald.
Chapter 1. Ellis Horwood, 1982.

1   Handicapping Weightlifters

1.1   Problem

Weightlifting is a minority sport, but it is a sport which can be understood
readily by the layman. There are nine official bodyweight classes (see Table 3.1)
and there are two principal lifts, namely the *snatch* and the *jerk*. In the snatch,
the weight is pulled from the floor to a locked arm overhead position in a *single*
move, although the lifter is allowed to move or squat under the weight as it is
being lifted, whereas for the jerk, *two* movements are allowed, the first to the
chest and the second to the overhead position.

10   Table 3.1 lists the world weightlifting records, current on 31 July 1977. The
maximum bodyweight for each class is also shown.

Table 3.1

Bodyweight Class		Lift	Lifted	Name and Country	Date
	(kg)		(kg)		
Flyweight	(52)	S	109	A. Voronin (USSR)	18 Mar 1977
		J	141	A. Voronin (USSR)	18 Jul 1976
Bantamweight	(56)	S	120.5	K. Miki (Japan)	25 Oct 1976
		J	151	M. Nassiri (Iran)	2 Aug 1973
Featherweight	(60)	S	130	G. Todorov (Bulgaria)	25 May 1976
		J	161.5	N. Kolesnikov (USSR)	20 Jul 1976
Lightweight	(67.5)	S	141.5	A. Aibazian (USSR)	15 Jul 1977
		J	180	S. Pevzner (USSR)	16 Jul 1977
Middleweight	(75)	S	157.5	Y. Vardanyan (USSR)	7 May 1977
		J	195	V. Milotosyan (USSR)	30 Jan 1977
Light-Heavyweight	(82.5)	S	170	B. Blagoyev (Bulgaria)	25 May 1976
		J	207.5	R. Milser (W. Germany)	8 Apr 1976
Middle-Heavyweight	(90)	S	180	D. Rigert (USSR)	14 May 1976
		J	221	D. Rigert (USSR)	14 May 1976
Heavyweight	(110)	S	185	V. Khristov (Bulgaria)	10 Apr 1976
		J	237.5	V. Khristov (Bulgaria)	22 Sep 1975
Super-Heavyweight	(over (110)	S	200	K. Plachkov (Bulgaria)	25 May 1976
		J	255	V. Alexeev (USSR)	27 Jul 1976

S – snatch, J – jerk

Table 3.2

Bodyweight Class	Total Winning Lifts		
	S	J	TOTAL
Flyweight	105	137½	242½
Bantamweight	117½	145	262½
Featherweight	125	160	285
Lightweight	135	172½	307½
Middleweight	145	190	335
Light-heavyweight	162½	202½	365
Middle-heavyweight	170	212½	382½
Heavyweight	175	225	400

It is clear that as bodyweight increases so does the lift, but what is not clear is the functional form of this increase. The problem is not just a physics/mathematics problem, but is a real practical one, as for many competitions (e.g. school, college) there are just not sufficient numbers to warrant the utilisation of the nine weight classes. It is also of some interest, even in serious competitions, to have an 'overall' winner. So the problem is to design some form of handicapping, which compensates for the various bodyweights.

    Having formulated a method of handicapping weightlifters, use it to find an *overall* winner for the winning snatch lifts at the 1976 Montreal Olympic Games, which are detailed in Table 3.2.

1.2 Mathematical models for weightlifting
There have been many models formulated which have been used for handicapping weightlifters. We will firstly describe several of these models, and then in section 1.3 use them to answer the question (and compare the results).

(i) *Linear model*
In Fig 3.15, we have illustrated the world record data points from Table 3.1 for both the snatch and the jerk, assuming that lifters are at their maximum allowed weight for their class. It is evident that a straight line is not a very accurate model, but nevertheless it does have some use. For example, the television programme "TV Superstars" uses the model

$$L' = L - B ,$$

where L' is the handicapped lift, L the lift, and B the lifter's bodyweight. This model implies a linear relationship between L and B with a unit slope, and a check with the data points in Fig 3.15 indicates that this approximation is not very accurate. At the level of performance of superstars it turns out to be much worse, favouring the lighter competitors substantially. On the other hand, the model has as its virtue *simplicity*. It can be understood by the average television viewer, and for this purpose can be viewed as a 'good' model. Entertainment is the object of TV Superstars, and a more complicated form of handicapping might leave viewers in some doubt as to what was happening.

40

(ii) *Power law models*
Looking at the data points in Fig 3.15 suggests that some power law model should be a more accurate description than a linear model. So if we assume that $L = kB^\alpha$ where k and α are constants, then

$$\log L = \log k + \alpha \log B \; ;$$

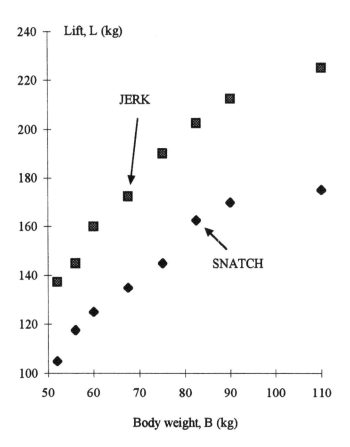

Fig 3.15

this means that a log L – log B graph should be a straight line, with slope α. For the world record data given above we obtain the points illustrated in Fig 3.16. A straight line looks a reasonable approximation for all the data points except
50   the heavyweight. Remembering, however, that the super-heavyweight data have not been included, a power law model does not seem entirely satisfactory, particularly for the upper weight classes. In fact, two power laws have been used extensively. These are as follows.

(a) *Austin formula* which is a 3/4 power law,

i.e. $L' = L/B^{3/4}$ (1.2)

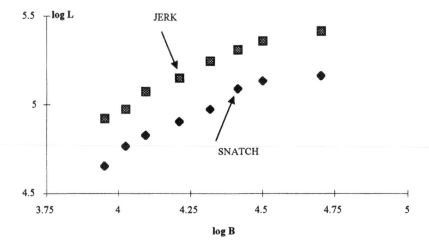

Fig 3.16

and was used some 20 years ago in the UK for odd lift comparison and championship best awards.

(b) *Classical formula* which is a 2/3 power law,

i.e. $L' = L/B^{2/3}$ (1.3)

60   and has been used for some time. This model can in fact be deduced from physiological arguments in simple terms. Briefly we assume that the lift is proportional to the average cross-sectional area of the lifter's muscle, say A,

i.e. $L = k_1 A$.   (1.4)

We also assume that A is proportional to the square of a typical body length, say l,

i.e. $A = k_2 l^2$,   (1.5)

and finally, the 'typical' body length is related to the bodyweight by

$B = k_3 l^3$.   (1.6)

With these assumptions,

70

$$L = k_1 k_2 l^2 = k_1 k_2 (B/k_3)^{2/3} = k\, B^{2/3} .$$ (1.7)

(iii) *O'Carroll formula*

The arguments presented above for the classical 2/3 power law are not altogether convincing. In 1967, an improved formula, designed by O'Carroll (1967) was introduced, and it has since gained wide acceptance by the weightlifting authorities. It is based both on zoological arguments and statistical analysis of the top 50 worldwide lifters for each division, overall time up to 1964. This was the most 'robust' and broadest based sample available, and it was used with an automatic compensation for the differing populations of competitors in the various bodyweight classes.

80   The assumptions (1.4), (1.5), and (1.6) above are generalised to:

$$L = K_1 A^x, x < 1 ,$$ (1.8)

which allows for the loss of efficiency at large size;

$$A = K_2 l^y, y < 2 ,$$ (1.9)

which allows for variations in shape. Here l is now a muscular length scale and

$$B - B_0 = K_3 l^3 .$$ (1.10)

This treats the total bodyweight $B = B_0 + B_1$ as the sum of B_0, being non-muscular weight such as cerebral which is almost independent of size, and $B_1 = K_3 l^3$ denoting the varying muscular and related weights. Now from (1.8), (1.9), and (1.10) we deduce a formula of the form

90      $$L = K(B - B_0)^\beta ,$$

where K and β are constants. From the statistical analysis in O'Carroll (1967), the values of the parameters are taken as

$$B_0 = 35 \text{ kg}, \quad \beta = 1/3 ,$$

giving the model as

$$L = K(B - 35)^{1/3} .$$

(1.11)

Using this model, the handicapped lift is

$$L' = \frac{L}{(B - 35)^{1/3}} .$$

(1.12)

In Fig 3.17 we illustrate a plot of $\log L$ against $\log(B - 25)$ for the data given in the introduction. We also illustrate straight lines with slope 1/3, which indicate the accuracy of this model. The three best performing classes (middle, light-heavy, and mid-heavy) are relatively highly populated with competitors, whereas the extreme bodyweight classes have relatively few competitors. This is probably the major factor behind the differences between the population-compensated formula and the population-affected world record figures.

100

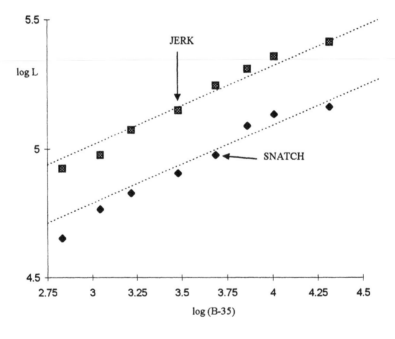

Fig 3.17

(iv) *Vorobyev formula*

This is a Russian formula which is based on determining a 'degree of merit' index *n*, which is given by

$$n = \frac{L+B}{B[0.45-(B-60)/900]}.$$ (1.13)

Note that $L + B$ is based on the idea that the lifter lifts himself along with the
110   barbell, so that the factor $(L + B)/B$ is the total lift as a multiple of bodyweight. The factor $[0.45 - (B - 60)/900]$ represents the proportion of available usable weight, which decreases with body weight.

The degrees of merit values for the world record lifts are given in Table 3.3.

Table 3.3

Bodyweight class	Degrees of merit	
	Snatch	Jerk
Flyweight	6.747	8.088
Bantamweight	6.936	8.135
Featherweight	7.037	8.204
Lightweight	7.010	8.301
Middleweight	7.154	8.308
Light-Heavyweight	7.201	8.371
Middle-Heavyweight	7.199	8.293
Heavyweight	6.800	8.010

Using this handicapping model, the best snatch lifter is the light-heavyweight (Blagoyev of Bulgaria) and the best jerk lifter is the middleweight (Milotosyan of the USSR).

1.3 Solutions of problem

We now use the models to find the 'overall' winner for the snatch winning lifts in the 1976 Olympic Games.
120   To compare the various models of handicapping we must select a common base; we adapt the handicapped lifts so that $L' = L$ when $B = 75$, the middleweight lifter's maximum bodyweight. That is to say, L' is an equivalent lift made at middleweight. From equations (1.1), (1.2), (1.3), (1.12), and (1.13) we obtain the handicapped lifts

1. Superstars	$L' = L - B + 75$	(1.15)
2. Austin	$L' = L(75/B)^{3/4}$	(1.16)
3. Classical	$L' = L(75/B)^{2/3}$	(1.17)
4. O'Carroll	$L' = L[40/(B - 35)]^{1/3}$	(1.18)
5. Vorobyev	$L' = [29250(L + B)B(465 - B)] - 75$	(1.19)

130   The handicapped lifts are given below for their *snatch* lift together with their numerical positions (assuming each lifter is at his maximum allowed bodyweight).

Bodyweight class	Snatch lift	HANDICAPPED LIFTS				
		Superstars	Austin	Classical	O'Carroll	Vorobyev
Flyweight	105	128.0 8	138.2 7	134.0 8	139.7 8	138.8 7
Bantamweight	117.5	136.5 7	146.3 4	142.8 6	145.7 4	146.6 4
Featherweight	125	140.0 5	147.7 3	145.0 3	146.2 3	147.7 3
Lightweight	135	142.5 4	146.1 5	144.8 5	144.7 6	145.8 5
Middleweight	145	145.0 3	145.0 6	145.0 3	145.0 5	145.0 6
Light-heavyweight	162.5	147.7 2	151.3 1	152.5 1	153.5 1	152.1 1
Middle-heavyweight	170	150.0 1	148.3 2	150.5 2	152.9 2	150.3 2
Heavyweight	175	140.0 5	131.3 8	135.6 7	141.9 7	138.5 8

Perhaps what is most surprising about these handicapped lifts is their agreement. With the exception of the Superstars model, all the models give the same first three places. In fact the Austin and Vorobyev models have complete agreement in all the rankings, and for all but the heavyweight the actual values of L' are very close. So, on the evidence here, we conclude that the best snatcher among the Montreal Olympics champions was the light-heavyweight, V. Shery (USA).

1.4 Reference

O'Carroll, M. J., (1967), On the relation between strength and bodyweight, *Research in Physical Education*, **1**, 6.

Questions

(i)   Which body weight class has been omitted from Fig 3.15? Explain why.

(ii)   With reference to Table 3.1 and Fig 3.15, if only the first seven classes were included, would it still be true to say (line [29]) "It is evident that a straight line is not a very accurate model"?

For the "snatch" lift only, find the slopes of the best fit straight line for the cases.

(a)   the first seven classes are included
(b)   the first eight classes are included.

(Hint: use a graphics calculator, a graph plotter or graph paper to plot the first seven points, draw the regression line and find its equation. Use the equation to calculate y for the seven values of x and compare with the observed values of y. Use this comparison to make a judgement about the accuracy of the model. Then add the eighth point and repeat the exercise.)

(iii) Is there a contradiction between the claim in line [37] that the "TV Superstars" linear model favours the "lighter competitors substantially" and the data in lines [134]-[137]?

(iv) Bearing in mind the results of part (ii) suggest a modified linear "TV Superstars" handicapping model which is fairer to all the competitors. Explain your answer.

(v) For the "snatch" lift only, reproduce Fig 3.16 using a graph plotter or graph paper and find the slope of the best fit line for

(a) the first seven classes
(b) the first eight classes.

Do these values match more closely the Austin formula or the Classical formula?

(vi) Consider these two statements:-

line [49] - "A straight line looks a reasonable approximation for all the data points except the heavyweight",

line [29] - "It is evident that the straight line is not a very accurate model".

Is the author guilty of "selective reporting"? Explain your answer.

(vii) Explain how each of the formulae at lines [32], [55], [59], [97] are modified to give the formulae at lines [125]-[128].

[Were you able to obtain line [129] from the formulae at line [108]?]

(viii) Describe and comment on the modelling assumptions made in developing the Classical formula.

(ix) Explain why the author believes that the O'Carroll formula is an improvement on the Classical formula.

(x) Using the Austin formula, and the 1992 world record data given in the attached table, decide which body weight class has improved the most between 1977 and 1992 for the "Snatch".

World Weightlifting Records, January 1992

(Taken from World Weightlifters Magazine)

Bodyweight Class		Lift	Lifted	Name and Country	Date
	(kg)		(kg)		
Flyweight	(52)	S	120.5	Z. Zairong (Chn)	27 Sep 1991
		J	155.5	I. Ivan (Bul)	27 Sep 1991
Bantamweight	(56)	S	135.0	L. Shoubin (Chn)	28 Sep 1987
		J	171.0	T. Neno (Bul)	6 Sep 1987
Featherweight	(60)	S	152.5	S. Naim (Tur)	20 Sep 1988
		J	190.0	S. Naim (Tur)	20 Sep 1988
Lightweight	(67.5)	S	160.0	M. Israil (URS)	18 Sep 1989
		J	200.5	P. Mikhail (Bul)	8 Sep 1987
Middleweight	(75)	S	170.0	G. Angel (Bul)	11 Dec 1987
		J	215.5	V. Alexander (Bul)	5 Dec 1987
Light-Heavyweight	(82.5)	S	183.0	Z. Asen (Bul)	14 Dec 1986
		J	225.0	Z. Asen (Bul)	12 Nov 1986
Middle-Heavyweight	(90)	S	195.5	B. Blagoi (Bul)	1 May 1983
		J	235.0	K. Anatoli (URS)	29 Apr 1988
Heavyweight	(110)	S	210.0	Z. Yuri (URS)	27 Sep 1988
		J	250.5	Z. Yuri (URS)	30 Apr 1988
Super-Heavyweight	(over (110)	S	216	K. Antonio (URS)	13 Sep 1987
		J	266.0	T. Leonid (URS)	26 Nov 1988

S = snatch   **J** = jerk

CASE STUDY 6 Insulating a House

Extract from: Mathematical Modelling – A Source Book of Case Studies, edited by
I D Huntley and D J F James. Chapter 7, Insulating a House, John Berry, Plymouth
Polytechnic, Oxford University Press, 1990.

Tables and figures have been omitted from the line numbering system.

7.   Insulating a House

7.1   The Problem Statement

Heating a house, or a flat, is an expensive part of the weekly
budget. In recent years the cost of heating fuels, that is,
coal, gas, electricity, oil, has increased in price quite
5   considerably.

It is important that as much of the heat as possible is
retained within the dwelling. Heat energy can escape through

the walls, windows, roofspace and floor, and there are several
products on the market to try to reduce this heat loss e.g.

10   (i) the cavity between the walls can be filled with various
materials; polystyrene balls and a chemical called urea
formaldehyde are just two such materials; (ii) the windows can
be double glazed forming a thin cavity of air between the two
panes of glass.

15   Fig 3.18 shows the relative proportions of heat loss for an
average sized detached house.

Clearly if some of escaping heat can be retained within the
house then less fuel is needed to heat the house to a certain
level and hence fuel bills will be lower. The heat loss can

20   be reduced by INSULATION.

New houses have a ten centimetre thickness of insulating
material in the loft reducing the heat loss through the
roofspace quite considerably. However the walls usually have
an unfilled air cavity between them and the windows consist of

25   a single pane of glass of about 4-6mm thickness.

A house owner wanting to reduce the fuel bills can invest in
"cavity-wall insulation" and/or "double glazing".

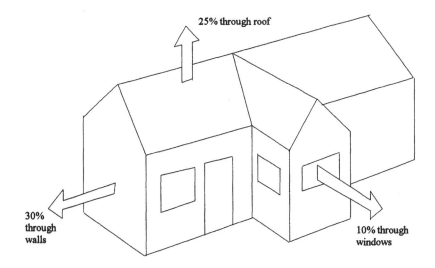

Fig 3.18   Heat losses from a house without roof or wall
insulation or double glazing. The remaining 35%
of heat loss goes through the floor, gaps round
the windows and doors, etc.

Figs 3.19 and 3.20 show portions of two advertisements for such forms of insulation.

30   Fig 3.19 suggests that cavity wall insulation could save nearly five times the amount than double glazing. But that soft of claim might be expected from a firm advertising the installation of cavity wall insulation. How good are these figures? This question is one that this modelling case study

35   sets out to answer. It is perhaps interesting to note that the double glazing advertisement only compares competing manufacturers and not other methods of insulation. Fig 3.20 introduces the term "U-value" which is one way in which heat losses through material can be defined.

40   Compare the relative cost effectiveness of installing cavity-wall insulation against double glazing. If you had to choose one form of insulation, by comparing the return on the capital investment, which would you choose? What factors that have not be included in your mathematical model, might affect your

45   decision? Would this change your decision?

This case study is ideally suited to students who have studied some basic physics or applied mathematics that has included some models of heat transfer. It shows how a simple linear model representing heat flow through a material can be used

50   to solve a problem of practical importance.

If students have not the background knowledge, then the relevant material is contained in this chapter.

7.2   Formulating a Model

7.2.1   Getting started

55   It is very tempting for this problem, to rush in to the physics of heat transfer and to start data collecting on the average temperatures both inside and outside the house. However that approach would soon lead to too much data and progress would seem slow. At some stage the process of heat flow through the

60   walls and windows does need investigating but first it is a good idea to stand back from the problem and see where we are going.

The problem statement asks for a comparison of the costs of installing two forms of insulation – double glazing and

65   cavity-wall insulation. So there are two aspects of the problem to consider:

DOUBLE
GLAZE YOUR
WALLS

Cavity walls, found in most houses built since 1920 were designed to keep damp from seeping in. Unfortunately, with today's central heating, they don't stop costly heat from seeping out. 35% of all heat lost, in fact. Only 10% escapes through the windows.

Cavity wall insulation still keeps damp out, but also keeps that heat in.

It can actually save about a

HOW MUCH MONEY YOU CAN SAVE FOR EVERY POUND SPENT ON FUEL	
25p	WITH CAVITY WALL INSULATION
16p	WITH LOFT INSULATION
7p	WITH FLOOR INSULATION
5p	WITH DRAUGHT EXCLUDER
5p	WITH DOUBLE GLAZING

Figures from Department of Environment

quarter of your heating bills (double glazing saves just 5%) and can pay for itself in about 5 years. So before you think about double-glazing, find out more about cavity wall insulation.

Surveys and quotations are FREE WITHOUT OBLIGATION. Phone, or Freepost the coupon, today for full details – You've nothing to lose.

PRIORITY ADVICE SERVICES Ring (0342) 123456

Fig 3.19 An advertisement for cavity wall insulation

The lower the U value the ore efficient the insulation.
The test results, shown in the chart below, speak for themselves.

	DOUBLE GLAZING AIR GAP	FRAME MATERIAL	U VALUE	INSULATION EFFICIENCY % INDEX 100
Single Glazed window		Aluminium	5.6	100%
Type A	20mm	Aluminium with plastic Thermal Break	2.2	254%
Type B	20mm	Aluminium with Plastic Thermal Break	2.9	193%
Type C	12mm	Aluminium with Plastic Thermal Break	3.0	187%
Type D	12mm	Aluminium with Plastic Thermal Break	3.2	175%
Type E	20mm	Aluminium with Plastic Thermal Break	3.3	170%
Type F	9.5mm	Aluminium	3.7	151%
		Type F also manufacture a Thermal Cad window		

Fig 3.20 An advertisement for double glazing

(i) the economics of the costs involved and

(ii) the physics of heat transfer through materials.

70 Formulating a mathematical model for the first aspect does not require any specialist knowledge of economics. However the model for heat transfer may either be a familiar one to the students or may require some investigation. In the description of possible solution to this problem a standard model is used.

75 The first stage is to write down a list of those features which we may need to take into account in the mathematical modelling process.

Perhaps the construction of a feature list would be a good place for the students to begin in a class discussion.

80 The following lists of features are relevant to the two different aspects of the problem already identified.

(i) Features affecting the 'economics'

1. cost of installation
2. cost of borrowing the money (i.e. interest paid)
85 3. cost of fuel
4. amount saved by insulating
5. type of double glazing
6. inflation

(ii) Features affecting the 'heat loss'

90 7. temperature of room
8. temperature of the outside
9. convection
10. conduction
11. radiation
95 12. area of walls
13. area of windows
14. thickness of glass
15. thermal properties of glass
16. thickness of walls
100 17. thermal properties of walls
18. heat saved by insulation

It is perhaps apparent that there are other features that do not seem to fit in the two categories given above. Two features that seem difficult to quantify are

105   (a)   comfort

      i.e.   double glazed windows are more comfortable to sit by
          than single glazed windows;

     (b)   attractiveness to the eye

      i.e.   "replacement windows" are possibly nicer to have in
110   the living area than "secondary double glazing".

The second one, (b), will affect the price of having the
house double glazed; so that if we choose double glazing by
installing replacement windows then, since this costs more
than secondary double glazing, a decision to have cavity-wall
115 insulation instead of double glazing may be because of the
concern for what it looks like.

7.2.2 Choosing variables and finding relations

The class could now be set the task of formulating a model
based on the feature lists given above. Each aspect, i.e.
120 the economics and the physics, can initially be investigated
separately although features 3 and 4 do provide a link.

With the list of 18 features given under headings (i) and (ii)
above, a model, which is based on the most important of them,
can now be formulated. Initially, for a simple model, we
125 will assume that the important economic features are

    the cost of installing double glazing = C_G

    the cost of installing cavity-wall insulation = C_B

Features 3 and 4 provide the link between the two aspects of
'economics' and 'heat transfer'. In fact

130
Amount of money saved = quantity of heat saved ×	I
in unit time cost of a unit amount of heat	

At this stage we will not worry about the 'type of double
glazing'. This feature affects the cost of installation so
that when a model to compare cost-effectiveness is formulated,
135 then, for different types of installation, different strategies
can be predicted.

Now consider the features that affect heat loss.

Fig 3.21 shows the physical processes involved with the heat transfer through walls and windows. The variable T

140 represents the temperature (in °C) and Q is the heat loss per unit time (measured in watts).

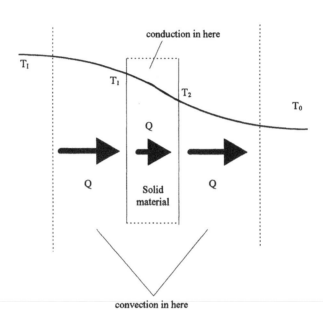

conduction in here

T_I

T_1

T_2

T_0

Q

Q

Solid
material

Q

convection in here

Fig 3.21

In Fig 3.21 the heat transfer due to radiation has not been considered. It is assumed in this model that the heat loss due to radiation is negligible in comparison with the heat loss due

145 to conduction and convection. Convection is due to the movement of the air on either side of the material causing temperature drops $T_I - T_1$ and $T_2 - T_0$. Conduction is due to contact between the particles making up the material and there is a temperature drop of $T_1 - T_2$ across its faces. The thermal properties of materials are modelled by a simple expression

150 called the U-value. This is a coefficient that relates the heat transfer per unit area to the total temperature difference ($T_I - T_0$ in the variables of Fig 3.21). The U-value can be calculated experimentally for different substances so that it takes into account convection and conduction. Fig 3.22

155 shows the U-values for walls and windows.

Material	U-value ($Wm^{-2}C^{-1}$)
Brick wall (no cavity)	1.92
Brick wall (with cavity not filled)	0.873
Brick wall (with filled cavity)	0.5
Single glass pane (6 mm)	6.41
Double glazed window	1.27

Fig 3.22 Typical U values for walls and windows

The basic heat transfer model is

| heat loss per unit area = U × (total temperature difference) | II |

This formula can be deduced theoretically by combining
160 simple models for convection and conduction, in the following
way.

Consider again the slab of material of thickness a separating
two regions of temperature T_I and T_0 shown in Fig 3.21.

Heating is conducted through the material because there is a
165 temperature difference $T_1 - T_0$ across its faces. The rate of
heat loss per unit area, Q, is related to $T_1 - T_2$ by the simple
linear model

$$Q = \frac{K}{a}(T_1 - T_2)$$

where K is the thermal conductivity and a is the thickness of
the material.

170 Convection occurs in thin boundary layers on each side of the
material. Simple linear models of convection for the inner
and outer boundary layers are given by

$$Q = h_1(T_I - T_1)$$

$$Q = h_2(T_2 - T_0)$$

175   where h_1 and h_2 are constants called convective heat transfer
coefficients. Their values depend on (a) the type of boundary;
(b) air speeds near the boundary. In each of the models for
conduction and convection we are assuming that the temperature
is independent of time, i.e. steady state conditions prevail.

180   Now we have three equations relating the temperatures T_I, T_1,
T_2 and T_0. They are

$$Q = h_1(T_I - T_1)$$

$$Q = \frac{K}{a}(T_1 - T_2)$$

$$Q = h_2(T_2 - T_0)$$

185   Eliminating T_1 and T_2 between these three equations we have

$$\left[\frac{1}{h_1} + \frac{a}{K} + \frac{1}{h_2}\right]Q = T_I - T_0$$

or $Q = U(T_I - T_0)$

$$\text{where } U = \left[\frac{1}{h_1} + \frac{a}{K} + \frac{1}{h_2}\right]^{-1}$$

is the U-value introduced earlier.

190   Typical values for a single glazed window are

$h_1 = 10 \text{ Wm}^{-2}\text{C}^{-1}$, $h_2 = 20 \text{ Wm}^{-2}\text{C}^{-1}$, $K = 1 \text{ Wm}^{-1}\text{C}^{-1}$ and $a = 0.006\text{m}$,

$$\text{giving } U = \left[\frac{1}{10} + 0.006 + \frac{1}{20}\right]^{-1} = 6.4 \text{ Wm}^{-2}\text{C}^{-1}.$$

Most of the features in list (ii) have been considered in this
simple model of heat transfer. Feature 18 can be written in
195   terms of the difference between the heat losses for non-
insulated and insulated boundaries.

We have

Heat saved by insulation	=	heat loss when uninsulated	−	heat loss when insulated		III

In terms of variables, we define

200 Area of glass $= A_G$ (same for single and double glazed windows)

Area of external walls $= A_B$ (same for non-insulated and insulated walls)

Amount of money saved per unit time $= \begin{cases} S_G \text{ for glass} \\ S_B \text{ for walls} \end{cases}$

Heat saved by insulation per unit time $= \begin{cases} H_G \text{ for glass} \\ H_B \text{ for walls} \end{cases}$

Cost of heat per unit time $= c$

205 Cost of double glazing $= \begin{cases} C_G \text{ for windows} \\ C_B \text{ for walls} \end{cases}$

Putting these variables into the three models I, II and III gives

<u>For the windows:</u>

II and III: $H_G = U_N A_G (T_I - T_0) - U_I A_G (T_I - T_0)$ (in Watts)

210 I: $S_G = c H_G$

where U_N and U_I are the U values for single and double glazed windows respectively.

<u>For the walls:</u>

II and III: $H_B = U'_N A_B (T_I - T_0) - U'_I A_B (T_I - T_0)$ (in watts)

215 I: $S_B = c H_B$

where U'_N and U'_I are the U values for unfilled and filled cavity walls respectively.

7.3 A Strategy for Cost-Effectiveness

220 Now we must make a decision as to the cost-effectiveness of installing double glazing or cavity wall insulation. Defining the "payback period" P, for each type of insulation to be the ratio of the cost of the insulation to the money saved due to

installing the insulation, then for the windows

$$P_G = \frac{C_G}{S_G}$$

225   and for the walls

$$P_B = \frac{C_B}{S_B}$$

A model for cost-effectiveness can then be formulated in the following way:

> If $\dfrac{P_G}{P_B} > 1$ then install cavity wall insulation
>
> If $\dfrac{P_G}{P_B} < 1$ then install double glazing

7.4 Putting it all Together

We have

$$\frac{P_G}{P_B} = \frac{C_G}{S_G} \times \frac{S_B}{C_B}$$

$$= \frac{C_G c(U'_N - U'_I)A_B (T_I - T_0)}{C_B c(U_N - U_I)A_G (T_I - T_0)}$$

$$= \frac{C_G A_B (U'_N - U'_I)}{C_B A_G (U_N - U_I)} = 0.0726 \frac{C_G}{C_B} \frac{A_B}{A_G}$$

235   If we use the U-values given in Fig 3.22, we have reached an
equation that does not include the data on temperature
differences and the fuel costs. Hence the increase of the
price of fuel due to inflation and other factors do not affect
the decision. Furthermore, the strategy that we adopt will
240   hold for all types of heating fuel.

To proceed further, suppose the cost of installing double
glazing and cavity wall insulation are calculated at a basic
rate per square metre c_G and c_B respectively, then

$$c_G = c_G A_G$$

245    $$c_B = c_B A_B$$

so that

$$\frac{P_G}{P_B} = 0.0726 \frac{c_G}{c_B} = 0.0726x$$

where

$$x = \frac{c_G}{c_B}.$$

250   Fig 3.23 shows a graph of $\dfrac{P_G}{P_B}$ against x.

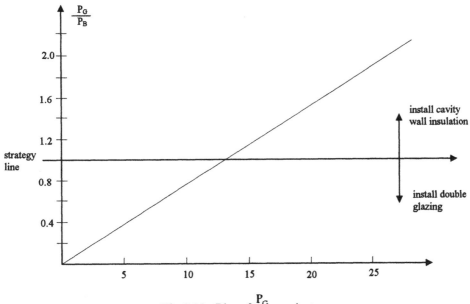

Fig 3.23 Plot of $\dfrac{P_G}{P_B}$ against x

From this graph we can make a decision whether to install
cavity wall insulation or double glazing. The value of X will
depend on the type of double glazing and cavity wall insulation,
and probably the manufacturer.

255   **7.5 Making a Decision**

Table 3.5 gives some examples of the cost per square metre of
cavity wall insulation and double glazing for the author's
house.

Table 3.5

Type of insulation	Average cost per square metre in pounds (Winter (1981)
Professionally fitted sealed double glazed units replacement windows	181
Professionally fitted secondary units	28.9
Do-it-yourself secondary double glazing (frame and single pane)	16
Do-it-yourself sealed units	30
Cavity wall insulation	2

Using this data and the graph in Fig 3.22 we have the ratio
$\frac{P_G}{P_B}$ for the four double glazing systems given by 6.57, 1.05,
0.58, 1.09 so that only for do-it-yourself secondary double glazing does the payback period become less than for cavity wall insulation.

7.6 Towards a More Complicated Model?

There are several features that may change the decision made on purely economic grounds. For instance it is more comfortable sitting by a window that has been double glazed.

If comfort is considered as important as cost then perhaps we should consider a model based on double glazing the "living" rooms only; or alternatively we could add a weighting so that for instance, we would install double glazing provided

$$\frac{P_G}{P_B} < 2 \text{ say instead of 1.}$$

We have not taken into account the heat saved by lining the curtains and drawing them more often. Curtains can save quite a lot of heat although there is still an uncomfortable draught under and around the edges of the curtains.

Installing cavity wall insulation and double glazing can increase the value of the house. An improved model could include an amount that takes this into account so that the true cost of the installation is reduced by a "recoverable amount".

280

Exercise 1

Which of the double glazing systems becomes cost effective (explaining what you mean by this) if we adopt a strategy of

(a) $\dfrac{P_G}{P_B} < 2$

(b) $\dfrac{P_G}{P_B} < 3$?

Exercise 2 Perfect double glazing

The following diagram shows ways in which heat is lost through a double glazed window and the convective and conduction coefficients.

Using the simple models of heat loss introduced in the text, i.e. $Q = \dfrac{K}{a}(T_x - T_y)$ for conduction and $Q = h(T_x - T_y)$ for convection show that the theoretical formula for the U value of a double glazed window is

$$U = \left[\frac{1}{h_1} + \frac{2K}{a} + \frac{1}{h_c} + \frac{1}{h_2} \right]^{-1}$$

Using the values given in the diagram calculate a value for U.

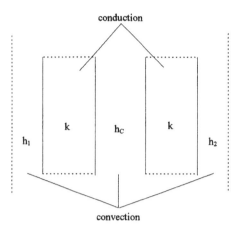

$h_1 = 10 \quad h_2 = 20 \quad h_c = 1.6 \text{ (in Wm}^{-2}\text{C}^{-1})$
$k = 1 \text{ wc}^{-1}$
$a = 6 \text{ mm}$

Diagram for Exercise 2

Exercise 3

The U value calculated in Exercise 2 is independent of the size of the double glazing air gap, b. However the advertisement shown in Fig 3.20 would suggest that the air gap size is important. Give reasons why this is so and how could the air gap size be taken into account?

Exercise 4

The calculation for perfect double glazing in Exercise 2 gives a U value much less than the quoted values in Fig 3.22. Can you give reasons to explain this apparent difference between the mathematical model and the experimental values?

Questions

(i)   In line [1] it is claimed that "heating a house or flat is an expensive part of the weekly budget". State whether you agree or disagree with this statement, giving reasons for your answer. Include appropriate facts and figures.

(ii)   Referring to line [8], are there any other routes for heat escape?

(iii)   Referring to Fig 3.18, is the greatest heat loss per unit area through the walls, the roof or the windows? Estimate areas for these routes.

(iv)   Referring to Fig 3.19 and your answer to question (i), estimate the cost of cavity wall insulation. Does this concur with your other enquiries/experiences or the data in Table 3.4? Explain your answer.

(v)   Describe briefly the difference between "replacement windows" and "secondary double glazing". (Refer to lines [108]-[110]).

(vi)   Define "U-value". (Refer to line [157]).

(vii)   List the assumptions made in creating the heat transfer model. (Refer to lines [136] to [189]).

(viii) Comment on the statement in line [205].

(ix)   In line [221] why does it make sense to call P the "payback period"? Explain how P could be defined differently to take account of the hidden economic costs (e.g. cost of borrowing the capital, or loss of interest that might otherwise have been earned).

(x)   Indicate how the value 0.0726 is obtained in line [234].

(xi)   Correct the misprints in lines [244], [245].

(xii)   Why do you think $h_2 > h_1$?
        Why do you think $h_c \ll h_1$ and h_2?
        (Refer to the diagram for Exercise 2)

(xiii) Do Exercise 2.

(xiv)   Do Exercise 4.

(xv)   Which feature of double glazing does most to reduce the U-value?
        (Refer to question (xiv).

4 •Developing Modelling Skills

In the previous two chapters we have illustrated modelling in action by studying population growth models and through published reports of modelling activities. In this chapter you have further opportunities to develop modelling skills by applying the ideas introduced earlier.

Section 4.1 provides a range of modelling problems to try. Sections 4.2 and 4.3 look at two specific types of modelling approaches that are important in business/economic problems and in science. Section 4.2 introduces **simulation modelling** and section 4.3 shows how **dimensional analysis modelling** can provide simple algebraic models in science and engineering.

4.1 Modelling Problems

CASE STUDY 1   Route Lengths

When planning a journey in advance one of the factors that might be considered important is the LENGTH of the route to be taken. To obtain a rough estimate of the length of the route the following model for use in conjunction with ordnance survey maps has been proposed:

> "Count the number of grid squares which the route enters, and then multiply this number by an appropriate factor to give the length of the route".

Investigate whether there is in fact any relationship between the length of a route and the number of grid squares it enters, and if appropriate suggest a factor that could be used. As your investigation develops you may feel it appropriate to suggest an alternative model.

CASE STUDY 2   Car Breakdown Relay Service

A new car breakdown service (similar to that offered by AA or RAC) is to operate a Relay Service by dividing the country into a number of regions. A broken-down vehicle is towed to the boundary of the towing vehicle's region. There it is transferred to another towing vehicle which tows it across the next region to the next boundary. This process continues until the home region of the broken-down vehicle is reached. It has been suggested that this procedure is more efficient than using one tow home for the whole journey.

Formulate models for such relay services and use your models to propose the best strategy.

CASE STUDY 3   Traffic Calming in a Village

The residents of the small village of Crapstone, near Plymouth in Devon, are concerned about the dangers from cars which ignore the 30 mph speed limit through the village. They are planning to approach West Devon council to install traffic calming measures. A sketch map of Crapstone is shown in Fig 4.1. The speed limit signs of 30 mph are placed at points marked *A* and *B*.

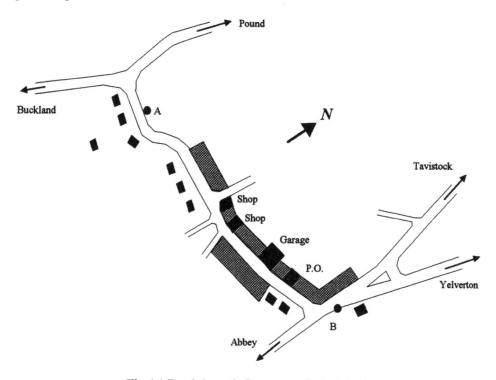

Fig 4.1 Road through Crapstone: Scale 1:4000

A common way of attempting to discourage motorists from driving at excessive speeds is the installation of speed humps (often called 'sleeping policemen').

By formulating a suitable mathematical model, make recommendations on how many speed humps are needed and how far apart they should be placed to calm the traffic passing through Crapstone.

CASE STUDY 4   Lorry Lanes

If you drive along the A38 into Cornwall, you will notice that on several of the hills there are 'crawler lanes'. These lanes are designed for slow moving vehicles to travel up the hill at their own pace without holding up faster moving vehicles. Consider the effect of driving uphill on lorries and recommend a policy for the provision of crawler lanes.

CASE STUDY 5 An Ordering Policy for a Bookshop

As the owner of a bookshop you wish to have a policy for ordering books which can be implemented by your storeman. Given that at least one copy of a book should be ordered how will he determine the number of copies to be ordered from the distributor? Some distributors will refund a fixed portion of the cost price of a book if the retailer returns it within a specified time and in mint condition, but if not it may still pay to have a few copies of a slow moving book on the shelves to cater for those who buy on impulse. Develop a model of the ordering procedures of your bookshop, and thus derive an appropriate policy.

CASE STUDY 6 The Milkman's Walk!

Your local dairy has decided to issue instructions for its milkmen, and it wants to include advice about the distance a milk truck should be driven between stops.

Develop a mathematical model to analyse different methods of delivery, and advise the dairy about the alternatives and how many houses the milkman should deliver to on foot at each stop.

CASE STUDY 7 Epidemics

Part 1 The Spread of a Cold

About every six weeks a ship calls at the isolated island of Tristan da Cunha. Sometimes one of the islanders catches a cold from a sailor and it spreads through the island community. Over a period of several years the islanders were asked to make a note of the day on which a cold developed. The data below come from one particular epidemic.

Day	New Cases	Day	New Cases
1	1	11	3
2	1	12	4
3	1	13	2
4	0	14	2
5	8	15	1
6	8	16	0
7	15	17	0
8	4	18	1
9	23	19	0
10	5	20	0

Table 4.1

Develop a model of the spread of a cold through the community.

Part 2   A general model for epidemics

The Local Health Authority is concerned at the rapid spread of infectious diseases (such as influenza). Although they cannot prevent a particular outbreak they are interested in possible control through vaccination. Also employment officers are interested in predicting how many people are likely to catch a particular disease and also how many people are likely to be off work as a result.

You are called in to help by developing a mathematical model of the spread of disease. Since the Authority may be asked for information about a number of different illnesses, the model is to be constructed in general terms as follows:

Consider the total population to be divided into the following three categories at any moment:

SUS: number of susceptible individuals at risk
INF: number of infected virus carriers
CUR: number of cured (and now immune) individuals.

Assume that

(a)   the percentage of susceptible individuals who contract the disease every day is proportional to the number of infected people;

(b)   a constant percentage of infected people recover from the disease every day.

Write down differential equations for SUS, INF and CUR and use them to study the spread and eventual subsidence of the epidemic starting with particular initial values of the three variables and using various values of the percentages mentioned in the assumptions.

If a vaccine giving 100% immunity is available but very costly and difficult to administer, investigate the effects of different vaccination policies.

CASE STUDY 8   Spread of Rabies

Rabies, a viral infection of the central nervous system, is transmitted by direct contact. Only about four people die of the disease in Europe in each year but the disease is much more common among animals. The European epidemic has now almost reached the north coast of France and it seems increasingly possible that rabies will reach Britain. The main carrier of the rabies virus is the red fox and with the high urban and rural fox population and the large number of domestic cats and dogs in England, the disease could spread particularly rapidly.

A mathematical model is required to help give information about the spatial spread of rabies, to predict how quickly the epidemic will spread, to estimate the number of rabid foxes at any given time and to examine the effects of various possible control strategies. For these purposes the fox population can be divided into three groups:

1. susceptible foxes,
2. infected foxes that are not yet infectious,
3. infectious, i.e. rabid foxes.

Develop a mathematical model based on the data given together with any assumptions you consider appropriate.

Use the model to study the progress of an epidemic starting from a point in the south of England.

Two possible means of containing the epidemic are:

(a) create a rabies 'break' by killing foxes in the areas adjacent to the infected area;

(b) vaccinate foxes (chicken heads infused with the vaccine can be left around for the foxes to eat).

Use the model to investigate and compare the possible effectiveness of these strategies.

Further Information

1. The density of the fox population varies from place to place but the average is about 4 foxes/km^2. Foxes tend to stay within a certain area, 'their territory' and do not stray into neighbouring territories.

2. Rabies is transmitted by direct contact, usually by biting and susceptible foxes become infected at a rate proportional to the rabid population.

3. During the infected period the animal seems to act quite normally. The average incubation time is about 28 days although there are documented cases of greater than 6 months.

4. The rabid stage lasts from 1 to 10 days. If the virus enters the spinal chord the rabid fox is paralysed, whereas if it enters the central nervous system the rabid animal exhibits the typical aggressive symptoms associated with the disease, the so-called 'furious rabies'. About half of the rabid foxes get the 'furious' form.

5. Rabid foxes almost always die. The mean life expectancy of a rabid fox is roughly 5 days.

6. Foxes with 'furious rabies' become aggressive, lose their territorial behaviour and wander randomly. This random wandering is important in spreading the disease.

CASE STUDY 9   Parachute Jumping

How long can a parachutist free fall before opening his or her parachute, if a comfortable landing is to be made? Consider jumps made from a moving aeroplane or a static balloon.

Develop a mathematical model to describe the motion of a parachutist and determine when the parachute should be opened for different starting conditions.

CASE STUDY 10   Soothing Babies

Many pregnant mothers have observed that while they are moving, or walking about, their unborn babies lie very still, but that when they rest, their babies become much more active. After babies have been born a rocking action will often comfort and send restless babies to sleep. It is suspected that there is a link between the motion of the unborn babies and the rocking that has a comforting effect when the babies have been born.

Investigate the motion that an unborn baby is likely to experience while the mother is walking and use your results to recommend how to comfort young babies.

4.2 Simulation Modelling

So far in this book we have formulated mathematical models in terms of algebraic symbolic models. In many problem solving situations the situation under investigation cannot be easily modelled analytically in this way or the relevant data required to enable us to model the system cannot be collected. In such circumstances, we can attempt to model the system using a **simulation model**. For example:

(i)   When designing a new aircraft, the design engineer must investigate the air flow over the aircraft. This could be done using the theory of fluid dynamics. It is much easier, however, to *simulate* the problem by building a scale model of the aircraft and by investigating the behaviour of the model in a wind tunnel.

(ii)   When investigating the queues that form in a bank - with a view to determining the numbers of cashiers needed at different times - the bank manager could experiment with different numbers of cashiers at different times until he came up with a pattern that worked. It is much simpler, however - and less trying on the patience of staff and customers - to *simulate* the possibilities using a computer and suitable data.

We shall concentrate on the second sort of simulation, particularly on the simulation of queues.

This section is merely an introduction to this modelling approach. A background knowledge of probability and statistics is needed to formulate anything other than the simplest simulation models. We begin with a case study.

Case Study: A queuing problem at an airport
(Source: Open University Course M371 Block IV Unit 1)

The activities of a large international airport can involve many queues. There are queues at the checking-in desks; there are queues at the customs and passport control desks; and there are queues of aircraft waiting to take-off or land. In this subsection we shall look at just one of the queues in detail – the queue of aircraft waiting to land.

Quite often at large international airports, incoming flights have to wait before they can land because all the runways are currently in use. Waiting is usually achieved by circling near the airport until a runway becomes free. Under normal circumstances, if several aircraft are waiting to land, they are allowed to land in the order in which they arrive. For the present we shall ignore the effect of aircraft taking off and any ground activity at the airport, and shall simply concentrate on the arrivals. We shall consider a *single* runway for the exclusive use of incoming flights. The situation is shown in Fig 4.2.

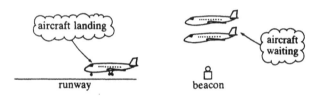

Fig 4.2 Aircraft landing and waiting to land at a single runway

When an aircraft arrives over a beacon it can do one of two things: (i) it can start landing, or (ii) it can wait over the beacon. (When waiting over the beacon, the aircraft joins a queue circling above the airport). If there are no aircraft waiting and the runway is free, then the aircraft can start to land; otherwise it must wait.

Consider a situation involving the arrival of four aircraft over the beacon. Table 4.2 shows the time between the arrival of successive aircraft over the beacon (the inter-arrival time) and the time it takes each aircraft to land (the landing time). The landing time is measured from the beginning of the approach to the runway from the waiting position until the aircraft is completely clear of the runway, when the runway is declared free.

	Aircraft 1	Aircraft 2	Aircraft 3	Aircraft 4
Inter-arrival time (seconds)	46	22	214	245
Landing time (seconds)	270	222	52	26

Table 4.2 A time-line diagram of the arrival and landing of aircraft on a single runway

The first aircraft arrives 46 seconds after the observations began, and we shall assume that the runway is clear so that it can start its approach. The time for Aircraft 1 to complete the landing procedure and free the runway is 270 seconds, so that the next aircraft cannot commence to land until a time $46 + 270 = 316$ seconds after the observations began. While Aircraft 1 is landing, Aircraft 2 arrives on the scene, at time $46 + 22 = 68$ seconds after the observations began, and thus has to queue for 248 $(= 316 - 68)$ seconds. Aircraft 2 can then begin its descent to the runway. The situation is illustrated in the **time-line diagram** in Fig 4.3.

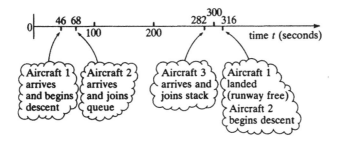

Fig 4.3 A time-line diagram of the arrival and landing of aircraft on a single runway

Aircraft 3 arrives at time $t = (46 + 22) + 214 = 282$ and joins the queue with Aircraft 2, which doesn't start its descent until time $t = 316$.

TUTORIAL PROBLEM 1

How long will Aircraft 3 have to wait? What happens to Aircraft 4?

Clearly, as more and more aircraft arrive, the situation can become quite complex. However, we can keep track of what is going on by noting that there are two points in time when something significant happens to the system. The first is when an arrival occurs and the second is when a landing is completed. These points in time are called **events**, and are important for two reasons. The first is that we only need to observe the system at each event, since the **state of the system** does not change between events. At any given time, a system may be described by the values of a number of discrete variables, known collectively as the **state of the system** at that time. The values of the variables are known as **states** and are such that they do not change between **events**, so

that the state of the system does not change between events. The second is that the
length of the queue of waiting aircraft only changes at an event. The events summarised
in Table 4.2 are illustrated in the time-line diagram in Fig 4.4.

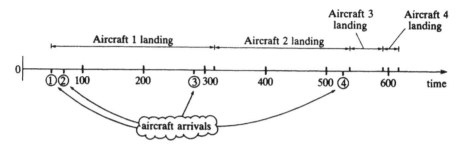

Fig 4.4 The time-line diagram of the events summarised in Table 4.2

For the system in Fig 4.4, the state of the system at any particular time can be described
in terms of four variables – one for each aircraft – each of which can take the four
discrete values or **states**: 'not arrived', 'waiting', 'landing' and 'landed'. The state of the
system at any particular time can be deduced easily from Fig 4.5. For example, the state
of the system at time $t = 400$ can be described as Aircraft 1 'landed', Aircraft 2 'landing',
Aircraft 3 'waiting', Aircraft 4 'not arrived'.

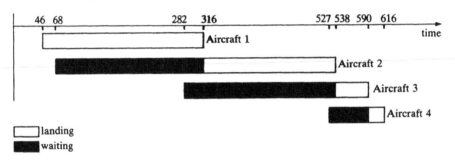

Fig 4.5 The waiting and landing times for the four aircraft

We are now in a position to retabulate the information in Table 4.2 in terms of events, as
in Table 4.3. You will notice that we simply list the various events, together with
relevant information about the system at that event. In other words, we simply update
our information about the system as the result of an event happening, and we effectively
ignore the system between events. Provided we include every relevant occurrence as an
event, the table will contain all the information we need about the system.

It is common in queuing theory to include the 'customer' being 'served' in the number in
a queue. So, for example, if at a bank there is just one person being served and nobody
waiting, then we say that there is a queue of one. If someone else arrives and stands
behind that person, waiting to be served, then we say that the length of the queue is two.
In Table 4.3, therefore, the length of the queue includes the aircraft landing as well as
the aircraft waiting. Only when an aircraft has completed its landing do we consider it
to have left the queue. In light of this, you should note that the number of aircraft
actually waiting to land is one less than the length of the queue.

Event	Time from start	Length of queue	Runway in use	Next arrival time	End of current landing	Total time runway not in use
0 Initial state	0	0	no	46	–	0
1 Arrival (1)	46	1	yes	68	316	46
2 Arrival (2)	68	2	yes	282	316	46
3 Arrival (3)	282	3	yes	527	316	46
4 End landing (1)	316	2	yes	527	538	46
5 Arrival (4)	527	3	yes	–	538	46
6 End landing (2)	538	2	yes	–	590	46
7 End landing (3)	590	1	yes	–	616	46
8 End landing (4)	616	0	no	–	–	46

Table 4.3 The arrival and landing of four aircraft

So far we have examined a simple pattern of aircraft arrivals and landings and have developed a method of recording the relevant information. Now suppose that the airport authorities at our single-runway airport expect an increase in aircraft traffic in the next few years and have to decide whether to build an additional runway. One way of making the decision is to use the average waiting time of an aircraft as a measure of the need for an additional runway. The authorities could wait until the average waiting time is judged to be too long and then start to build the new runway. However, as the new runway could take several years to construct, adopting this approach could cause airlines to complain and choose to use another airport. It would be better for the airport authorities to plan ahead and try to predict just when the additional runway would be required. To do this, a mathematical model is needed which can predict the average waiting time for different arrival patterns of aircraft. The sort of mathematical model we shall use is a simulation model.

To provide a complete simulation of aircraft landing patterns, several features need to be considered:

(i)   the arrival pattern over the beacon;

(ii)   the landing times;

(iii)   the queue discipline (e.g. do the aircraft land in the order in which they arrive, or do perhaps long-haul flights get preference?);

(iv)   any initial conditions (e.g. is the runway clear when the first aircraft arrives?).

Let us consider the arrival pattern first. To simulate a series of arrivals, we need to know the typical arrival pattern of aircraft over the beacon. Suppose that Table 4.4 shows data collected for 200 successive aircraft arriving at the airport on a typical day.

In practice we would need to collect data on several days, as we could not be sure that any particular day was 'typical', but for the purposes of this example we will just use one day's data to produce a theoretical simulation of the arrival of aircraft.

The inter-arrival time is the time between the arrival of successive aircraft over the beacon. For example, the table tells us that on 44 occasions an aircraft arrived within 60 seconds of the aircraft in front of it.

Inter-arrival time (seconds)	Number of aircraft	Relative frequency	Cumulative frequency
0 – 59	44	0.22	0.22
60 – 119	34	0.17	0.39
120 – 179	27	0.135	0.525
180 – 239	22	0.11	0.635
240 – 299	16	0.08	0.715
300 – 359	13	0.065	0.780
360 – 419	10	0.05	0.830
420 – 479	8	0.04	0.870
480 – 539	6	0.03	0.900
540 – 599	5	0.025	0.925
600 – 659	4	0.02	0.945
660 – 719	3	0.015	0.960
720 – 779	2	0.01	0.970
780 – 839	2	0.01	0.980
840 – 899	1	0.005	0.985
900 – 959	1	0.005	0.990
960 – 1019	1	0.005	0.995
1020 – 1079	0	0	0.995
1080 – 1139	1	0.005	1
1140 – 1199	0	0	1

Table 4.4 Data on the arrival of 200 aircraft

The numbers in the *relative frequency* column are found by dividing the number of aircraft in each inter-arrival time interval by 200 (i.e. the total number of aircraft). The numbers in the *cumulative frequency* column are found by adding relative frequencies up to and including the corresponding inter-arrival time interval (e.g. the cumulative frequency for the inter-arrival time interval 120–179 is $0.22 + 0.17 + 0.135 = 0.525$). For the inter-arrival time interval 120–179, for example, the relative frequency of 0.135 tells us that 13.5% of the 200 aircraft have an inter-arrival time of between 120 and 179 seconds and the cumulative frequency of 0.525 tells us that 52.5% of the 200 aircraft have an inter-arrival time of less than 180 seconds.

The first stage in the simulation is to model a random arrival of an aircraft using the data. A graph of the cumulative frequency provides a means of choosing inter-arrival times from the typical day's data in Table 4.4. Fig 4.6 shows this graph, in which each point is joined to its neighbour by a straight line.

For any given cumulative frequency, we can read off from the graph the corresponding range of inter-arrival times. For example, a cumulative frequency of 0.43 corresponds to the straight-line segment of the graph joining time $t = 120$ to time $t = 180$, and so the corresponding inter-arrival time interval is 120–179. Thus to model a random arrival, we could take any random number from a table of random numbers and read the corresponding inter-arrival time interval from the graph.

To be of much use in a simulation of the airport's activities, we need a more accurate value for the inter-arrival time than, for example, 'between 120 and 179 seconds'. This can be done using *linear interpolation*. For example, consider the random number 0.43. Now, instead of looking for the corresponding line segment on the graph, we read off directly from the graph the time corresponding to the point on the line segment that corresponds to a cumulative frequency of 0.43. We thus get an inter-arrival time of approximately $t = 140$.

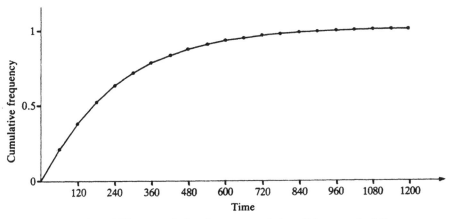

Fig 4.6 The cumulative frequency of aircraft inter-arrival times

Using the cumulative frequency graph to perform linear interpolation is rather tedious and not very accurate. In particular we would have had difficulty in using the graph in Fig 4.6 if our random number were given to more than two decimal places. Instead we use an algebraic approach, based on the properties of a straight line.

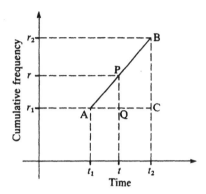

Fig 4.7 Part of a cumulative frequency graph

Consider Fig 4.7, which shows just one of the line segments of a cumulative frequency graph. A and B are the ends of the straight line segment, with coordinates (t_1, r_1) and (t_2, r_2). For example, A might be the point $(120, 0.39)$ and B the point $(180, 0.525)$ in Fig 4.7. From the formula for the gradient of the line AB, we have

$$\frac{r - r_1}{t - t_1} = \frac{r_2 - r_1}{t_2 - t_1}.$$

Rearranging to give a formula for t, we obtain

$$t = t_1 + \frac{r - r_1}{r_2 - r_1}(t_2 - t_1). \tag{1}$$

Hence given a random number r, we first find the segment of the cumulative frequency graph which corresponds to r and then we can calculate t using equation 1. For example, $r = 0.43$ corresponds to the third segment of the graph in Fig 4.6. Hence $t_1 = 120$, $r_1 = 0.39$, $t_2 = 180$ and $r_2 = 0.525$. Therefore the corresponding inter-arrival time, correct to the nearest second, is

$$t = 120 + \frac{0.43 - 0.39}{0.525 - 0.39}(180 - 120) = 138.$$

We have seen how we can model a single random arrival. It is now an easy matter to build up a possible profile of arrivals over a period of time by simply repeating the procedure of **sampling** from the graph of the cumulative frequency of inter-arrival times.

Example 1

By successive sampling from the graph of the cumulative frequency of inter-arrival times in Fig 4.6, build up a possible profile of five arrivals at the airport using the random numbers in Table 4.5.

0.17593
0.24106
0.42675
0.11392
0.50623
0.83532
0.35854
0.31886
0.70904
0.31016

Table 4.5 Random number table

SOLUTION

The first five random numbers from the table of random numbers are 0.17593, 0.24106, 0.42675, 0.11392, 0.50623. Using Fig 4.6, Equation 3 and rounding the results to the nearest second we get the inter-arrival times shown in Table 4.6.

random number	Inter-arrival time (seconds)
0.17593	48
0.24106	67
0.42675	136
0.11392	31
0.50623	172

Table 4.6 Sampled inter-arrival times for Example 1.2

TUTORIAL PROBLEM 2

Continue the sampling process of Example 1 for the next five random numbers in the table of random numbers.

We now have a simulation for the arrival of the first ten aircraft at the airport, based on arrival data for a typical day. In a similar way we can simulate the landing of these first ten aircraft by sampling from a cumulative frequency graph based on landing data for a typical day. The result of such a simulation is shown in Table 4.7.

Aircraft number	Landing time (seconds)
1	196
2	222
3	270
4	52
5	26
6	108
7	286
8	63
9	37
10	67

Table 4.7 Simulated landing times

Using the simulated inter-arrival times and landing times, we are now nearly in a position to simulate aircraft landings for the ten aircraft arrivals, by carrying out the same procedure as we did for the arrival of the four aircraft earlier in this subsection. All we need is information on the queue discipline and on the initial conditions. We shall simply assume that the aircraft land in the order in which they arrive and that initially (at time $t = 0$) the runway is free and there are no aircraft waiting to land.

TUTORIAL PROBLEM 3

Using the simulated data in Table 4.6, in your solution to Tutorial Problem 2 and in Table 4.7, and assuming that aircraft land in the order in which they arrive and that at time $t = 0$ the runway is free and no aircraft are waiting to land, simulate the arrival and landing of ten aircraft by setting up a table similar to Table 4.3. Use your table to determine:

(i)   the time at which the tenth arrival occurs;

(ii)   the maximum number of aircraft waiting to land at any one time;

(iii)   the percentage of the total time during which the runway was in use up to the time when the tenth landing was completed.

We have seen how we can simulate the pattern of aircraft arrival and landing based on data derived from the current pattern. If you recall, the problem we were trying to solve was to determine whether, and if so when, to build an additional runway. To do this we need to simulate the future patterns of arrivals and landings. We can perform such a simulation using theoretical cumulative frequency graphs that model the expected patterns of inter-arrival times and landing times. But this is beyond the purpose of this section; our aims have been to introduce the idea of simulation as a method of modelling.

TUTORIAL PROBLEM 4

Suppose that customers arrive at a cafeteria, which has a single serving bay, after one of two inter-arrival times – 20 seconds and 80 seconds; and that the service time (i.e. the time taken to be served) is either 15 seconds or 65 seconds. Assume that each inter-arrival time and each service time is equally likely to occur.

(a) By tossing a coin, and letting a head represent the arrival of a customer 20 seconds after the previous one and a tail represent the arrival of a customer 80 seconds after the previous one, draw up a table showing the arrival of the first 20 customers of the day. Similarly use a coin to model the service times for the first 20 customers.

(b) Assuming that customers are served in the order in which they arrive, use the data from (a) to perform a simulation of the arrival and service of the first 20 customers of the day.

(c) What is the maximum number of customers queuing at the service bay and which customer has the longest wait between arriving and receiving his or her food?

Simulation Modelling Problems

CASE STUDY 11 Dentist Appointments

Devise a simulation model to examine the following appointment system used by a dentist:

1. Patients are given appointments every 10 minutes from 9.00 am to 12 noon.

2. 10% of patients fail to turn up.

3. Some patients arrive early, some late. The distribution is:

	early		on time	late	
	10 min	5 min		5 min	10 min
Probability	0.1	0.2	0.4	0.2	0.1

4.   Consultation times have the following distribution:

Time spent with dentist (mins)	5	10	15	20	25	30	35
Probability	0.1	0.15	0.2	0.25	0.15	0.1	0.05

5.   Patients are seen in the order in which they arrive.

Discuss how you might use your model to investigate the efficiency of the appointment system.

CASE STUDY 12   Bank Queues

The manager of a bank is considering changing over to a new single queue system but is not sure if it will be better for customers than the existing system.

In the present system there are five service points and when customers enter the bank they can choose to queue at any of the five points. If a customer at the back of any queue notices that another cashier has become free he will move over for service. The average time between the arrival of customers in the busy period is 0.5 minutes and the average time taken to serve a customer is 2.5 minutes.

Devise a simulation model which will enable you to compare the two systems and help the bank manager to make his decision.

CASE STUDY 13   Supermarket Cashouts

A small supermarket has four cash tills and the time taken to serve a customer at any of the tills is proportional to the number of items in his/her trolley (roughly one second per item). Twenty percent of customers pay by cheque or credit card which takes 1.5 minutes, paying with cash takes only 0.5 minutes.

It is proposed to make one cash till a quick service till for customers with 8 items or less. Two of the remaining tills are to be designated "cash only".

Develop a simulation model which will enable you to compare the operation of the proposed system with that of the present system.

CASE STUDY 14   Queuing for a Lift

From time to time queues can develop for lifts in office blocks and hotels. It is common for lifts to visit all floors collecting passengers on the way up and down. However in some buildings there are other arrangements, such as 'express lifts' that miss out floors.

Formulate a simulation model to investigate a range of strategies for scheduling lifts.

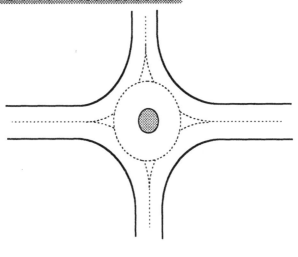

Fig 4.8

The sketch shows a traffic roundabout with four arms, each a single carriageway. When the traffic flow rate is low any vehicle is able to proceed through the roundabout without delay. Above some critical flow rate q, queues may form as a result of drivers waiting at the give-way lines for acceptable gaps between vehicles already circulating in the roundabout.

Develop a mathematical model to enable you to investigate how the critical flow rate q depends on the other parameters defining the roundabout. In particular, what is the relationship between q and the circulating flow rate Q on the roundabout?

Your model should also give statistical data on queuing times at the entry lanes.

Can you extend your model to 3 and 5 arm roundabouts?
Also consider entries with 2 and 3 lanes instead of just one.

4.3 Dimensional Analysis Modelling

The essence of mathematical modelling is to formulate a mathematical model that adequately represents a given situation. We begin by choosing relevant variables to represent the important features in the problem and seek mathematical relationships connecting the variables. In this section we introduce a fundamental property of a physical quantity called its **dimensions** and investigate a method of using dimensions to find effective groupings of the variables. This is called **dimensional analysis modelling** and is often used in forming relationships in science and technology.

Units and Dimensions

All of the physical quantities that you have met so far can be expressed as the 'amount or size' of the quantity followed by a unit of measurement. For example, the acceleration due to gravity is often written as 9.8ms^{-2} and the force of gravity on an object of mass 10 kg is 98 kgms^{-2} or 98N. In most courses in science you will be familiar with the international system of units called SI units. These have a 'base' unit system and other units are built from them. For example, we could use the unit kgms^{-2} for force showing the mass × acceleration law for forces. However this is cumbersome and the unit of force adopted is the Newton so that 1 kgms^{-2} = 1N.

Of course we do tend to mix up units, for example, the speed of a car is often quoted in miles per hour, so that miles and hours are base units, and you might give your mass in terms of pounds (or stones and pounds) instead of kilograms.

It is sometimes convenient to separate physical quantities from their units. For example, a speed is always a length divided by a time whether we use miles per hour, metres per second or furlongs per year. So we can think of length and time as fundamental quantities.

In science and engineering we choose the three quantities mass, length and time and define them as **base dimensions** denoting them by M, L and T.

> the dimension of mass is M
> the dimension of length is L
> the dimension of time is T

The dimensions of other physical quantities are then written in terms of M, L and T. In order to write this more succinctly we use square brackets [ ] to mean "the dimension of". For example,

> [mass] = M, [time] = T.

Example 2

Find the dimensions of speed and area using (a) their SI units, and (b) appropriate formulae for each quantity.

SOLUTION

(a)   The SI unit of speed is ms^{-1}, a unit of length divided by a unit of time, so we can write

$$[\text{speed}] = \text{LT}^{-1}.$$

The SI unit of area is m^2, the product of two units of length, so we can write

$$[\text{area}] = L^2 .$$

(b)   Alternatively we can use formulae which define speed and area. For an object travelling at constant speed we have

$$\text{speed} = \frac{\text{distance}}{\text{time}} ,$$

so that $[\text{speed}] = \dfrac{L}{T} = LT^{-1} .$

For area we know that for a rectangle

$$\text{area} = \text{length} \times \text{breadth} ,$$

so that $[\text{area}] = L \times L = L^2 .$

TUTORIAL PROBLEM 5

Find the dimensions of the following quantities using (a) their SI units, and (b) appropriate formulas for each quantity

    acceleration, volume, force, density, pressure, energy, angle.

The dimensions of angle may have given you some discussion. It is an example of a **dimensionless quantity**. The angle measured in radians is defined from a circle by $\theta = l / r$ (see Fig 4.9) where l is the arc length and r is the radius.

$$[\text{angle}] = \left[\frac{l}{r}\right] = \left[\frac{L}{L}\right] = L^0 = 1$$

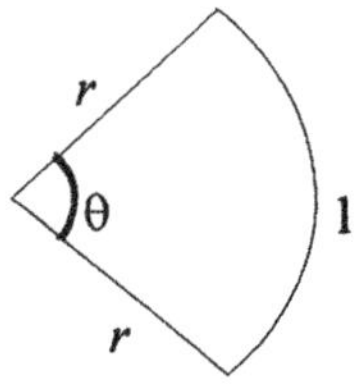

Fig 4.9

A **dimensionless quantity** is one whose dimensions are $M^0 L^0 T^0$ or 1.

The numbers at the start of formulas are often dimensionless quantities. For example, the area of a circle is πr^2 and the π is dimensionless; the kinetic energy of an object of mass m, travelling with speed v is $\frac{1}{2}mv^2$ and the number $\frac{1}{2}$ is dimensionless.

The coefficient of friction, μ, for a block sliding on an inclined plane is an example of a dimensionless constant. We can show this from the definition of the friction force as $F = \mu N$ where N is the normal reaction,

$$[\mu] = \frac{[F]}{[N]} = \frac{MLT^{-2}}{MLT^{-2}} = M^0 L^0 T^0 = 1 .$$

The dimensions of mass, length and time are sufficient to express the dimensions of physical quantities when temperature does not change. For problems involving temperature change we need to introduce a fourth base dimension of temperature, denoted by the symbol Θ.

Example 3

The equation of state for a gas is

$$p = R\rho T$$

where p is pressure, ρ is density, T is temperature and R is a constant. What are the dimensions of R?

SOLUTION

$$R = \frac{p}{\rho T}$$

$$[R] = \frac{[p]}{[\rho][T]} = \frac{ML^{-1}T^{-2}}{ML^{-3}\Theta} = L^2 T^{-2} \Theta^{-1}$$

You will notice from this example that some physical constants have dimensions (e.g. R) and some are dimensionless (e.g. μ). Clearly care is needed.

Dimensional Consistency

In any equation or formula we expect each side to have the same units, so that the dimensions must be the same on each side. We say that the equation or formula is **dimensionally consistent** (sometimes **dimensionally homogeneous**). Any statement that is dimensionally inconsistent **must** be wrong.

Example 4

Show that the formula, with the usual convention,

$$s = ut + \tfrac{1}{2}gt^2$$

is dimensionally consistent.

SOLUTION

We need to look at the dimensions of each term in the equation

$$[s] = L$$

$$[ut] = LT^{-1}T = L$$

$$[\tfrac{1}{2}at^2] = LT^{-2}T^2 = L$$

All three terms have the same dimensions so the equation is dimensionally consistent.

TUTORIAL PROBLEM 6

Which of the following equations are dimensionally inconsistent?

(a) $\quad v^2 = u^2 + 2as$

(b) $\quad \tfrac{1}{2}mv^2 = mgh + mv$

(c) $\quad F = pV$

(d) $\quad p + \tfrac{1}{2}\rho u^2 = mgs$

(e) $\quad F = \mu mv$

(f) $\quad v = u - gt$

where v and u are speeds, a and g are accelerations, s and h are distances, m is mass, F is force, p is pressure, V is volume, ρ is density, t is time and μ is the coefficient of friction.

The fact that an equation or formula is dimensionally consistent does not imply that it makes sense physically. For example, $v^2 = u^2 + a^2t^2$ is dimensionally consistent but is not a mathematical model for any real life physical situation.

The mathematical modelling of the physical world makes sense only if the models formulated are dimensionally correct. Good modellers develop the habit of checking all the models they formulate for dimensional consistency. This is one reason for obtaining symbolic models avoiding the temptation to substitute for constants. For example, it is better to leave the kinetic energy of an object of mass 10kg moving with speed v as $K = \tfrac{1}{2}mv^2$ rather than $5v^2$. In the former all the dimensions are apparent whereas for $K = 5v^2$ the dimensions are no longer evident.

Dimensional Modelling

The idea, introduced in the previous section, that each term in an equation must have the same dimensions can be used to formulate possible models between physical quantities. The process is called **dimensional modelling** or **dimensional analysis**.

We illustrate the method of approach with an example.

Example 5

A simple pendulum consists of a bob of mass m attached to a fixed point by a light inextensible string of length l. The bob performs small oscillations through a maximum angle θ. Find a formula for the period of oscillations t in terms of m, l, θ and the acceleration due to gravity g.

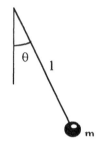

Fig 4.10 Simple pendulum

SOLUTION

If starting this problem without the question wording our first step would be to write a feature list and try to formulate a model from the important features. Such a list would include the features given in the problem description.

physical quantity	symbol	dimensions
time for one complete swing	t	T
length of pendulum	l	L
mass of bob	m	M
acceleration due to gravity	g	LT^{-2}
maximum angular deflection	θ	1

We are seeking a mathematical model of the form

$$t = f(l,m,g,\ \theta)\ .$$

Any relationship of this type must be dimensionally consistent so that each side must have dimensions T. We begin with the **assumption** of a power law:

$$t = kl^{\alpha}m^{\beta}g^{\gamma}\theta^{\delta}$$

where the powers α, β, γ, δ are to be found and k is a dimensionless constant. For this mathematical model to be dimensionally correct we have to find α, β, γ, δ such that

$$[t] = [kl^{\alpha}m^{\beta}g^{\gamma}\theta^{\delta}]\ .$$

In terms of M, L and T we have

$$T = [k]L^{\alpha}M^{\beta}(LT^{-2})^{\gamma}[\theta]^{\delta}$$

$$= L^{\alpha}M^{\beta}(LT^{-2})^{\gamma}$$

since $[\theta] = 1$ (angles are dimensionless) and $[k] = 1$ from the assumption. Hence

$$T = L^{\alpha+\gamma}M^{\beta}T^{-2\gamma}.$$

Equating powers of L, M and T gives

L: $\qquad \alpha + \gamma = 0$

M: $\qquad \beta = 0$

T: $\qquad -2\gamma = 1$

Solving these for α, β and γ gives

$$\gamma = -\tfrac{1}{2}, \quad \alpha = -\gamma = \tfrac{1}{2} \quad \text{and} \quad \beta = 0.$$

We conclude that a simple power law model is

$$t = kl^{\frac{1}{2}}g^{-\frac{1}{2}}\theta^{\delta} = k\sqrt{\frac{l}{g}}\theta^{\delta}.$$

The method tells us nothing about δ and the relationship with angle θ. Since k and θ are dimensionless quantities we replace $k\theta^{\delta}$ by some dimensionless function of θ, $k(\theta)$ say, giving

$$t = k(\theta)\sqrt{\frac{l}{g}}.$$

A dimensionless approach has given a model. It is at this stage that we would validate the model using suitable experimental data. In fact it turns out that for small angles θ, less than about $10°$, the period is approximately

$$t = 2\pi\sqrt{\frac{l}{g}}$$

so that $k(\theta) = 2\pi$, a constant. As the maximum angular displacement θ increases, the function $k(\theta)$ varies with θ and the model breaks down.

Dimensional modelling has not given us the exact formula for t in terms of l, g, m and θ, but it does predict two important results:

- the period is independent of the mass m,

- the quantity $\sqrt{\dfrac{l}{g}}$ is an important time scale in the problem.

This modelling approach gives an insight into how the variables are related to each other and suggests suitable experiments for investigation of the system.

Notice how the modelling process introduced in Chapters 1 and 2 is evident in the solution. We begin with identifying the important features and variables, formulate a model using dimensional consistency and validate the model with experimental data.

TUTORIAL PROBLEM 7   Laminar flow in a pipe

For the smooth laminar flow of water along a very long horizontal pipe the following features are thought to be important.

physical quantity	symbol	dimensions
pressure drop along the pipe	p	$ML^{-1}T^{-2}$
viscosity of water*	μ	$ML^{-1}T^{-1}$
diameter of the pipe	d	L
volume flow rate in a pipe	Q	L^3T^{-1}

* The viscosity is a measure of the stickiness of a fluid.

Use dimensional modelling to formulate a model for the pressure drop along the pipe.

Investigate possible relationships for an improved model which includes the length l of the pipe as an important feature.

TUTORIAL PROBLEM 8   Water waves on the ocean

An applied mathematician assumes that the speed of waves on the surface of an ocean, c, depends on the depth of the ocean, h, the density of the water, ρ, the acceleration due to gravity, g, and the wavelength of the wave, λ.

(a)   Initially she decides to ignore λ. Use dimensional modelling to formulate a model for the speed c_0. These are called shallow water waves because $h \ll \lambda$.

(b)   In the next model she decides to include λ but ignore the depth h. Use dimensional modelling to formulate a model for the speed c_∞. These are called deep water waves because $h \gg \lambda$.

TUTORIAL PROBLEM 9 A leaking tank

A water container is filled to a depth h. When a small hole is drilled in the bottom of the container it takes t seconds for the water to run out. It is assumed that t depends on h and the acceleration due to gravity g.

(a)   Use dimensional modelling to formulate a model for t.

(b)   Design an experiment to obtain appropriate data to validate your model and find a value for the unknown dimensionless constant.

In the example and problems considered so far there has always been one combination of the relevant variables chosen. The following extension of Tutorial Problem 7 shows that this is not always the case.

Example 6

The speed of water flowing through a horizontal pipe, v, depends on the following physical quantities.

physical quantity	symbol	dimensions
pressure drop	p	$ML^{-1}T^{-2}$
length of pipe	l	L
viscosity of water	μ	$ML^{-1}T^{-1}$
diameter of pipe	d	L
density of water	ρ	ML^{-3}

In forming this feature list the effects of the roughness of the pipe have been neglected and the flow is assumed to be laminar (i.e. smooth and not turbulent). (The viscosity is a measure of the stickiness of a liquid).

Use dimensional modelling to formulate a model for the pressure drop in terms of the listed features and the speed v.

SOLUTION

Assume a power law model of the form

$$p = kv^{\alpha} l^{\beta} \mu^{\gamma} d^{\delta} \rho^{\varepsilon}$$

where k is a dimensionless constant.

Dimensional consistency implies

$$[p] = [k][v]^\alpha [l]^\beta [\mu]^\gamma [d]^\delta [\rho]^\varepsilon$$

$$ML^{-1}T^{-2} = (LT^{-1})^\alpha (L)^\beta (ML^{-1}T^{-1})^\gamma (L)^\delta (ML^{-3})^\varepsilon$$

$$ML^{-1}T^{-2} = M^{\gamma+\varepsilon} L^{\alpha+\beta-\gamma+\delta-3\varepsilon} T^{-\alpha-\gamma}$$

Equating powers of M, L and T gives

M: $1 = \gamma + \varepsilon$

L: $-1 = \alpha + \beta - \gamma + \delta - 3\varepsilon$

T: $-2 = -\alpha - \gamma$

giving three equations in five unknowns. To proceed further we need to choose to write three of the variables in terms of the other two.

Suppose we let $\alpha = a$ and $\beta = b$, two unknown parameters. Solving for γ, δ and ε in terms of a and b

$$\gamma = 2 - \alpha = 2 - a$$

$$\varepsilon = 1 - \gamma = 1 - (2 - a) = a - 1$$

$$\delta = -1 - \alpha - \beta + \gamma + 3\varepsilon = -1 - a - b + 2 - a + 3(a-1) = a - b - 2$$

The power law model becomes

$$p = kv^a l^b \mu^{2-a} d^{a-b-2} \rho^{a-1} = k \frac{\mu^2}{\rho d^2} \left(\frac{vd\rho}{\mu}\right)^a \left(\frac{l}{d}\right)^b .$$

The quantities $\dfrac{vd\rho}{\mu}$ and $\dfrac{l}{d}$ are dimensionless and each choice of a and b gives a possible combination of the variables. So we write

$$p = \frac{\mu^2}{\rho d^2} f\left(\frac{vd\rho}{\mu}, \frac{l}{d}\right) .$$

The quantity $\dfrac{\mu^2}{\rho d^2}$ has dimensions of pressure so we think of this as a 'pressure scale' associated with the laminar flow of water in a pipe.

TUTORIAL PROBLEM 10

By choosing other pairs of the unknown powers α, β, γ, δ and ε find other pressure scales for the laminar flow of water in a pipe.

So the power law model leads to five 'pressure scales' in the problem of the flow of water through a pipe. It is at this stage of the modelling process we would need to interpret and validate the mathematical relationships through experimentation of physical laws. Dimensional modelling can provide the combinations of the relevant variables with the required dimension but it is physical laws that can help us to select the appropriate combination that best models the physical situation.

TUTORIAL PROBLEM 11 Simple pendulum with air resistance

The simple pendulum in Example 5 experiences an air resistance force F. Use dimensional modelling to find a model for the period of small oscillations of the pendulum.

TUTORIAL PROBLEM 12 Drag force on a sphere in a fluid

The drag force, F, on a smooth sphere dropping vertically through a liquid depends on the following physical features: speed of the sphere, v, diameter of the sphere, d, the liquid density, ρ, and the liquid viscosity μ. Use dimensional modelling to formulate a simple model for F.

Design an experiment to validate your model.

Answers to Selected Problems

Chapter 1

Tutorial Problem 4

$$\frac{R}{R_e} = 0.4 + 0.3 \times 2^n.$$

Chapter 2

Tutorial Problem 5

(a) $\qquad P_{n+1} - P_n = P_n - 0.05 P_n.$

where P_n is the size of the population of deer n years after May 1980. We assumed that half of the population is female and each female has 2 young so that the number of births is P_n per year.

$$P_{n+1} = 1.95 P_n.$$

(b)

Year (May)	1980	1981	1982	1983	1984	1985	1986	1987	1988	1989	1990
Size of herd	50	97	189	368	717	1398	2726	5315	10364	20209	39407

(c) 168 should be killed at the end of 1982 and then 190 should be killed each year.

Exercises

1. (a) $P_{n+1} = 0.8 P_n + 1000.$

    (b) Equilibrium population when $P_{n+1} - P_n = 0$ i.e. $P_n = 5000$ birds.
    For $P_0 = 3000$ population increases to 5000.
    For $P_0 = 8000$ population decreases to 5000.

2. (a) $P_{n+1} = 0.95 P_n + 500.$

    (b) Equilibrium population is 10000 fish.
    For $P_0 = 5000$ population increases to the equilibrium value.
    For $P_0 = 15000$ population decreases to the equilibrium value.

3. $P_{n+1} - P_n = (0.7 - 3 \times 10^{-5} P_n) P_n$.

Stable equilibrium is 2.33×10^4 birds.

With shooting of the birds: $P_{n+1} - P_n = (0.5 - 3 \times 10^{-5} P_n) P_n$.

New stable equilibrium is 1.67×10^4 birds.

The population of birds is within 1% of the equilibrium population after 7 years.

4. $P_{n+1} - P_n = (0.6 - 3 \times 10^{-4} P_n) P_n$.

Stable equilibrium is 2000 fish.

With restocking of the lake: $P_{n+1} - P_n = (0.8 - 3 \times 10^{-4} P_n) P_n$.

New stable equilibrium is 2666 fish.

The stock level is within 5% of this equilibrium population after 3 years.

Chapter 3

Case Study 1

Exercise 1

1.   60°. This is within the required range for θ.

Case Study 2

(ii)   From a graph we found $\alpha = 2$ and $\log_e k = -1.3$.

Then $k = 0.05$ and $S_B = 0.05V^2$.

(iii)   Equation (9.2) should read $S = 0.682V + 0.076V^2$.

(v)   The equation "$v^2 = u^2 + 2as$" follows from assuming that the particle moves with a constant acceleration a. The initial and final speeds are u and v, and s is the displacement. In our case, final speed $v = 0$, initial speed $u = V$ and displacement $s = S_B$.

Hence $V^2 = -2aS_B$.

But $S_B = 0.076V^2$ or $V^2 = 13.16 S_B$.

Comparing these equations lets us see that the car is moving with a constant negative acceleration (or deceleration) of magnitude 6.58 ms^{-2}.

The braking force required to produce this deceleration is 658N.

(viii) The smallest angle that can cause a stone to reach height h for a given V must be that angle that gives equal roots for t in part (vi),

i.e. $V^2 \sin^2 \theta = 2gh$ or $\sin\theta = \dfrac{\sqrt{2gh}}{V}$ or $\theta = \sin^{-1}\left(\dfrac{\sqrt{2gh}}{V}\right).$

(ix) Tractors do not have an overhanging boot and so stones are projected at angles greater than 35°. Thus, even when travelling at a speed and distance reckoned to be safe for cars, it is no longer safe; you are too close.

The advice to give is "leave more space".

(xi) Modify equation (9.1) to ensure that $S - Vt_E < kX_E$ where k is some factor > 1. You could try $k = 2$.

Follow this through the rest of the calculation.

Case Study 3

(i) $c = c_0 e^{-kt}$ where c_0 is a constant.

(ii)

Concentration (mg/l)	Time (hours)	$12e^{-0.17t}$ (to 1 decimal place)
10.0	1	10.1
7.0	3	7.2
5.0	5	5.1
3.5	7	3.7
2.5	9	2.6
2.0	11	1.8
1.5	13	1.3
1.0	15	0.9
0.7	17	0.7
0.5	19	0.5

(iii) $c(nT^+)$ is the concentration immediately after the injection given at time $t = nT$.

(v) Hint: $1 + e^{-kT} + e^{-2kT} + \ldots + e^{-nkT}$ is a geometric series with sum $\dfrac{(1 - e^{-knT})}{(1 - e^{-kT})}.$

(vii) (b) Minimum concentration = 5.1 mg/l.

(c) $\displaystyle\lim_{n \to \infty} c(nT^+) = 20.3$ mg/l; $\displaystyle\lim_{n \to \infty} c(nT) = 10.3$ mg/l

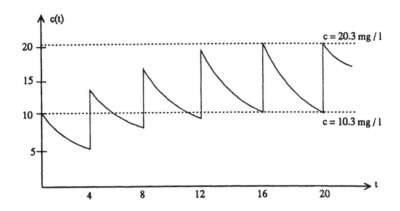

(viii) (a) $D = 450$ mg, $T = 3.5$ hours, $d = 200$ mg.

(b)

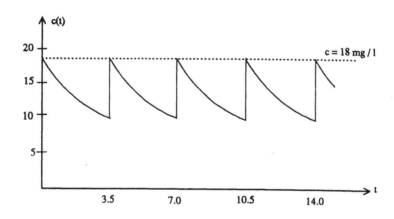

(ix) Strategy in (vii).

Advantages – case of administration every 4 hours is a convenient time interval and 250 mg is a convenient dose.

Disadvantages – concentration gets closer to the extremes of the therapeutic range thus cutting down on the margin for error in size of dose or time of injections.

Strategy in (viii).

Advantages – concentration is kept well within the therapeutic range.

Disadvantages – difficulty of administration since an injection every 3.5 hours is unlikely to fit in with the hospital routine – there is a larger initial dose.

Case Study 4

(i)   The author does not tell us, nor does he express an opinion. The likely reason is "overcrowding", but the author does say that "there was an abundance of food and places to live" (line 4).

(ii) $0.25 \times \text{acre} = 4840 \times 0.9144^2 / 4 = 1012\text{m}^2$.

Hence error = 1.2%

(iv)   When $k = 0.25$, $P = 192$ (from the graph). This is fairly close to 200.

(v)   He "fits" a straight line to the $k - P$ curve in the neighbourhood of $P = 200$.

Our best straight line (by eye) has equation

$$k = 0.004(260 - P).$$

If we follow this through into the differential equation we get

$$\frac{dP}{dt} = 0.2P\{0.004(260 - P) - 0.25\} = 0.001(200 - P)$$

which is the equation the author obtained.

The straight line $k = 0.005(250 - P)$ is not quite the same approximation to the curve, but leads to the same differential equation.

(vi)   For $P < 200$,

$$P = \frac{200}{1 + \left(Ae^{0.2t}\right)^{-1}}.$$

For $P > 200$,

$$P = \frac{200}{1 - \left(Ae^{0.2t}\right)^{-1}}.$$

(vii) $\dfrac{P(12)}{P_0} = \dfrac{200e^{12/5}}{200 + P_0\left(e^{12/5} - 1\right)}$.

(viii)

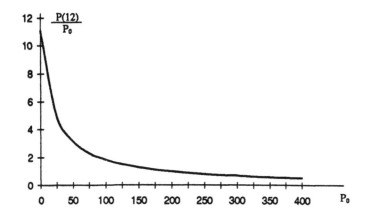

(ix) (a) $P_0 = 100$ gives $P(12)/P_0 = 1.8$.

(b) $P_0 = 200$ gives $P(12)/P_0 = 1.0$.

(c) $P_0 = 300$ gives $P(12)/P_0 = 0.7$.

(a) 80% growth in one year.

(b) no growth (equilibrium).

(c) reduction by 30% from 300 to 210 (death rate exceeds effective birth rate).

(x) Probably only (b) is valid. The graph in Fig 3.14, which we have used, is derived from a *linear* relationship only valid near $P = 200$. This straight line predicts a k value of 0.64 at $P = 100$, against a recorded value of 0.86. Also the predicted k value at $P = 300$ is negative whereas $k \geq 0$ for all P.

The author does not tell us what the value of P_0 was.

(xi) When a linear approximation is taken for the $k - P$ graph in Fig 3.13, we obtain the logistic equation in line 53. Hence the $k - P$ graph for the logistic equation is a straight line with negative slope like the equation in line 50.

(xii)   The modulus function comes in because

$$\int \frac{f'(x)}{f(x)} dx = \log|f(x)|$$

since $\log x$ is only defined for $x > 0$.

Case Study 5

(i)   The super-heavyweight class has been omitted because it does not have an upper class boundary.

(ii)   For the "snatch", with the first seven classes, the regression line is

$$L = 16.6 + 1.85B$$

and the (L,B) correlation coefficient is 0.996. This is a very good fit.

B	L(calc)	L(observed)	L(o) − L(c)
52	112.8	109	−3.8
56	120.2	120.5	+0.3
60	127.6	130	+2.4
67.5	141.5	141.5	0
75	155.3	157.5	+2.2
82.5	169.2	170	+0.8
90	183.1	180	−3.1

With the first 8 classes, the regression line is

$$L = 46.9 + 1.38B \qquad (r = 0.953).$$

This is not such a good fit as the seven-point line.

B	L(calc)	L(observed)	L(o) − L(c)
52	118.7	109	−9.67
56	124.2	120.5	−3.7
60	129.7	130	+0.3
67.5	140.1	141.5	+1.5
75	150.4	157.5	+7.1
82.5	160.75	170	+9.25
90	171.1	180	+8.9
110	198.7	185	−13.7

(iii)   The answer could be Yes or No!

Yes – The data in lines 1-4 of Table 4.4 clearly show that the lighter competitors fare badly under this handicapping system.

No – Since B is subtracted from L and B is smaller for the lighter classes, then it does favour these classes. The lighter classes come last simply because they are not very good.

(iv) A model based on the best-fit lines obtained in part (ii) is suggested, e.g.

$$L' = L - 1.5(B - 75),$$

where 1.5 is a compromise between 1.85 and 1.38. This provides the following:

B	L	Rank order	Austin rank
52	139.5	7	7
56	146	5	4
60	147.5	2 =	3
67.5	146.25	4	5
75	145	6	6
82.5	151.25	1	1
90	147.5	2 =	2
110	122.5	8	8

The ranking is consistent with more complicated models.

(v) Seven-point graph:

$$\log L = 1.169 + 0.898 \log B \qquad (r = 0.996).$$

Eight-point graph:

$$\log L = 1.825 + 0.740 \log B \qquad (r = 0.970).$$

This relates to $L = kB^a$ with $a = 0.90$ (7-pt), $a = 0.74$ (8-pt). This matches Austin ($a = 0.75$) more closely than classical ($a = 0.67$).

(vi) By now you should have noticed that the seven-point linear model is just as good a fit as the seven-point power-law model.

Since the authors reject the linear model because it is "not a very accurate model" but accept the seven-point power model as a "reasonable approximation", they are guilty of "selective reporting", i.e. choosing a result or prove *their* thesis (that the power law is better) and ignoring another result.

(vii) The requirement is to make $L' = L$ at $B = 75$, and so for the linear (additive) model a normalisation factor (constant) must be added. For the power-law models (multiplication), a normalisation factor must be multiplied.

We could not work out the relation between the index of merit n and the handicapped lift L'. If you take $n = L'$ in equation (1.13) then, by the same process as before (making $L' = L$ at $B = 75$) it is possible to obtain (1.19). There is a "divide" symbol "/" missing in (1.19).

(viii) (a) Lift is proportional to cross-sectional area of muscle. This seems a reasonable assumption. The lifting power of *n* "ropes" together is about *n* times the lifting power of one rope.

(b) Area is proportional to (body length)^2. This assumes a standard body shape and may not take account of variation in shape (e.g. tall thin people vs short fat people).

(c) Body weight is proportional to body volume and hence to (body length)^3 (same comment as at (b)).

(ix) The O'Carroll formula is based on a statistical analysis of a large amount of relevant data, and also on zoological arguments (which are not explained in the text). The formula allows for "the loss of efficiency at large size", and for "non-muscular" body weight.

For interest the seven-point O'Carroll best fit is

$$\log L = 3.50 + 0.42 \log(B - 35) \qquad r = 0.9988;$$

and the eight-point is

$$\log L = 3.64 + 0.38 \log(B - 35) \qquad r = 0.9878.$$

(x)

B	$L(92)$	$L'(92)$	$L(77)$	$L'(77)$	$\dfrac{L'(92) - L'(77)}{L'(77)}$	$\dfrac{L(92) - L(77)}{L(77)}$	$L'(92) - L'(77)$
52	120.5	158.6	109	143.5	0.11	0.11	15.1
56	135.0	168.1	120.5	150.0	0.12	0.12	18.1
60	152.5	180.3	130.0	153.7	0.17 (best)	0.17 (best)	26.6 (best)
67.5	160.0	173.2	141.5	153.1	0.13	0.13	20.1
75	170.0	170.0	157.5	157.5	0.079	0.079	12.5
82.5	183.0	170.4	170.0	158.3	0.076	0.076	12.1
90	195.5	170.5	180.0	157.0	0.086	0.086	13.5
110	210.0	157.6	185.0	138.8	0.14	0.14	18.8

In this case it does not matter whether one uses actual lifts or handicapped lifts. Provided one looks at relative improvement, the results are the same. The most improved is class 3 – Featherweight.

The formula $L'(92) - L'(77)$ produces the same winner but interchanges places 2 and 3 in the rank ordering (see table). Quite a different ordering is obtained by using $L(92) - L(77)$.

Case Study 6

(i)   Typical expenditure figures per month for a household are

Food	£400
Heating	£ 50
Mortgage	£250
Light	£ 30
Phone	£ 30
Car	£200
	£960

Of these major items of expenditure, heating is about 5%, which is not insignificant, it is relatively inexpensive for this household. (Or perhaps your experience is different ...?)

(ii)   Through the doors.

(iii)   One of the authors' (KH) house is approximately 10m × 10m × 8m high.
Each wall has area 10m × 8m = 80m^2.
Area of 4 walls is 320m^2.

I have 5 windows measuring 3m × 2m and 5 windows measuring 4m × 2m.
Total area 70m^2.

Hence area of (walls – windows) = 250m^2 = 62%
            area of windows   =   70m^2 = 13%
            area of floor of roof space = 100m^2 = 25%

Heat loss per unit area

walls	30% ÷ 250 = 0.0012
windows	10% ÷ 70 = 0.0014
roof	25% ÷ 100 = 0.0025

Hence heat loss through the roof is greatest.

(iv)   Heating costs per year = 12 × £50 = £600 (1991)

            25% of this = £150

and   over 5 years = £750.

According to Table 3.5, with insulation costs 250m^2 × £2 = £500 (but this figure relates to 1981 when the article was written. Perhaps costs have increased in 10 years).

(v)   Replacement windows are factory sealed double glazed units.
        Secondary double glazing is just another pane of glass (plus fittings) which is
        fitted inside an existing single glazed window.

(vi)   U = heat loss per unit time per unit area per degree of temperature gradient.
        (Did you remember to include "per unit time"?)

(vii)   line 142   radiation is negligible.

        line 168   uses a linear model for rate of heat loss per unit area which is
                           taken to be directly proportional to temperature differences and
                           inversely proportional to thickness of material.

        lines 173, 174   heat loss due to convection is proportional to temperature
                           difference across the convecting region.

        line 179   steady state conditions prevail i.e. temperature is independent of
                           time.

(viii) "Cost of double glazing" only refers to windows and not walls. A better variable
        name would be "cost of insulation".

(ix)   The units of P are "time"

$$P = \frac{C_g}{S_g} = \frac{\text{cost (money)}}{\text{money owed per unit time}} = \text{time}.$$

It is a measure of how long it takes for the savings to mount up to the original
capital cost.

To define P differently, replace C by $C*$ where $C*$ is the total cost of insulation,
including interest charges on money borrowed, or replace S by $S*$ where $S*$ is S
minus interest not earned on the capital per year, $S* = S - rC$.

(x) $$\frac{U'_N - U'_I}{U_N - U_I} = \frac{0.873 - 0.5}{6.41 - 1.27} \qquad \text{(using data in Fig 3.22)}$$

        $$= 0.0726$$

(xi)   There should be uppercase C letters on left-hand sides.

(xii) h depends on (line 177)

        (i)   type of boundary,
        (ii)   air speed near the boundary.

It is likely that air speeds on the outside of the window are greater than air speeds inside the house because of the wind.

h_c is much smaller than h_1 and h_2 because the air in the air-gap of the double glazed window is virtually static.

(xiii) $\quad U = \left\{ \dfrac{1}{h_1} + \dfrac{2a}{K} + \dfrac{1}{h_c} + \dfrac{1}{h_2} \right\}^{-1}$

There is a misprint in Exercise 2.

$$\dfrac{2K}{a} \text{ should be } \dfrac{2a}{K}.$$

(xiv) Using the correct expression for U

$$U = \{0.1 + 0.012 + 0.625 + 0.05\}^{-1}$$

$$= 1.27$$

which is the value in Fig 3.22. The misprint explains the difference the author refers to!

Using the incorrect expression

$$U = \{0.1 + 333.3 + 0.625 + 0.05\}^{-1} = 0.003 .$$

(xv) In the correct expression for U_1 the most significant term is 0.625 and this comes from the small amount of convection through the static air gap. So this feature does most to reduce the U value.

Chapter 4

Tutorial Problem 1

Aircraft 3 waits 256 seconds before it can commence landing.
Aircraft 4 arrives at $t = 527$, joins the queue and waits until $t = 590$ at which time the runway is free. Aircraft 4 completes its landing time in $t = 616$.

Tutorial Problem 2

random number	inter-arrival time (seconds)
0.835 32	428
0.358 54	109
0.318 86	95
0.709 04	296
0.310 16	92

Tutorial Problem 3

Event	Time from start	Length of queue	Runway in use	Next arrival time	End of current landing	Total time runway not in use
0 Initial state	0	0	no	48	–	0
1 Arrival (1)	48	1	yes	115	244	48
2 Arrival (2)	115	2	yes	251	244	48
3 End landing (1)	244	1	yes	251	466	48
4 Arrival (3)	251	2	yes	282	466	48
5 Arrival (4)	282	3	yes	454	466	48
6 Arrival (5)	454	4	yes	882	466	48
7 End landing (2)	466	3	yes	882	736	48
8 End landing (3)	736	2	yes	882	788	48
9 End landing (4)	788	1	yes	882	814	48
10 End landing (5)	814	0	no	882	–	48
11 Arrival (6)	882	1	yes	991	990	116
12 End landing (6)	990	0	no	991	–	116
13 Arrival (7)	991	1	yes	1086	1277	117
14 Arrival (8)	1086	2	yes	1382	1277	117
15 End landing (7)	1277	1	yes	1382	1340	117
16 End landing (8)	1340	0	no	1382	–	117
17 Arrival (9)	1382	1	yes	1474	1419	159
18 End landing (9)	1419	0	no	1474	–	159
19 Arrival (10)	1474	1	yes	–	1541	214
20 End landing (10)	1541	0	no	–	–	214

(i)   The tenth arrival occurs at time $t = 1474$ seconds.

(ii)   The maximum number of aircraft waiting to land at any one time is three (one less than the maximum length of the queue, which includes the aircraft actually landing).

(iii)   The runway was not in use for a total of 214 seconds, and was therefore in use for $1541 - 214 = 1327$ seconds, i.e. for 86% of the time.

Tutorial Problem 5

$$[\text{acceleration}] = LT^{-2}$$
$$[\text{volume}] \quad = L^{3}$$
$$[\text{force}] \quad = MLT^{-2}$$
$$[\text{density}] \quad = ML^{-3}$$
$$[\text{pressure}] \quad = ML^{-1}T^{-2}$$
$$[\text{energy}] \quad = ML^{2}T^{-2}$$
$$[\text{angle}] \quad = 1$$

Tutorial Problem 6

(b), (c), (d), (e) are dimensionally inconsistent.

Tutorial Problem 7

$$p = k\frac{\mu Q}{d^3}$$

If we include l,

$$p = \frac{\mu Q}{d^3}\,\text{f}\!\left(\frac{l}{d}\right); \quad p = \frac{\mu Q}{l^3}\,\text{f}\!\left(\frac{d}{l}\right)$$

Tutorial Problem 8

$$c_0 = k\sqrt{hg} \quad c_\infty = k\sqrt{\lambda g}$$

Tutorial Problem 9

$$t = k\sqrt{\frac{h}{g}}$$

Tutorial Problem 10

$$p = \frac{\mu^2}{\rho l^2}\,\text{f}\!\left(\frac{\rho vl}{\mu}, \frac{d}{l}\right); \quad p = \rho v^2 \text{f}\!\left(\frac{\mu}{\rho vl}, \frac{d}{l}\right); \quad p = \frac{\mu v}{l}\,\text{f}\!\left(\frac{\rho vl}{\mu}, \frac{d}{l}\right);$$

$$p = \rho v^2 \text{f}\!\left(\frac{\mu}{\rho vd}, \frac{l}{d}\right); \quad p = \frac{\mu v}{d}\,\text{f}\!\left(\frac{\rho vd}{\mu}, \frac{l}{d}\right)$$

Tutorial Problem 11

$$t = \sqrt{\frac{l}{g}} f\left(\frac{mg}{F}, \theta\right); \quad t = \sqrt{\frac{ml}{F}} f\left(\frac{mg}{F}, \theta\right)$$

Tutorial Problem 12

$$F = \rho d^2 v^2 f\left(\frac{\mu}{v\rho d}\right); \quad F = \frac{\mu^2}{\rho} f\left(\frac{vd\rho}{\mu}\right); \quad F = vd\mu f\left(\frac{vd\rho}{\mu}\right)$$

Index of Modelling Problems

Index

Printed and bound by CPI Group (UK) Ltd, Croydon, CR0 4YY

03/10/2024

01040331-0016